The

of
Colloid and Interface Science

The

of
Colloid and Interface Science

A Dictionary of Terms

Laurier L. Schramm

ACS Professional Reference Book

American Chemical Society   Washington, DC   1993

Library of Congress Cataloging-in-Publication Data

Schramm, Laurier Lincoln.
  The Language of Colloid and Interface Science: A Dictionary of Terms / Laurier L. Schramm.

    p.   cm.

  Includes bibliographical references.

  ISBN 0–8412–2709–8 (clothbound); ISBN 0–8412–2710–1 (paperback)

  1. Colloids—Dictionaries. 2. Surface chemistry—Dictionaries.
I. Title.

QD549.S37   1993
541.3'45'03—dc20   93–25680
                                                    CIP

1993 Advisory Board

About the Author

LAURIER L. SCHRAMM is senior staff research scientist and group leader for enhanced oil recovery process sweep improvement at the Petroleum Recovery Institute in Calgary. He is also an adjunct associate professor of chemistry at the University of Calgary, where he lectures in applied colloid and interface chemistry. He received his B.Sc. (Hons.) in chemistry from Carleton University in 1976 and Ph.D. in physical and colloid chemistry in 1980 from Dalhousie University, where he studied as a Killam  Scholar. From 1980 to 1988 he held research positions with Syncrude Canada Ltd. in its Edmonton Research Centre.

His research interests have included many aspects of colloid and interface science applied to the petroleum industry, including research into mechanisms of new processes for the enhanced recovery of light to heavy crude oils, such as foam, polymer, and surfactant flooding, and research into fundamental and applied aspects of the hot water flotation process for recovering bitumen from oil sands. These have involved investigations into the rheology of emulsions, suspensions, hydrocarbons, and oil foams; the electrokinetic properties of dispersed solids, oil, and gases in aqueous solutions; dynamic surface and interfacial tensions and phase attachments; and the reactions and interactions of surfactants in solution. Dr. Schramm is a member of the Chemical Institute of Canada (including serving on the Local Section Executive), the associations of the Chemical Profession of Alberta and Ontario, the American Chemical Society, and the International Association of Colloid and Interface Scientists. He has some 60 scientific publications and patents. This is his second ACS book.

Preface

In the early 1800s Thomas Graham studied the diffusion, osmotic pressure, and dialysis properties of a number of substances, including a variety of solutes dissolved in water[1]. He noticed that some substances diffused quite quickly through parchment paper and animal membranes and formed crystals when dried. Other substances diffused only very slowly if at all through the parchment or membranes and apparently did not form crystals when dried. Graham proposed that the former group of substances, which included simple salts, be termed crystalloids, and the latter group, which included albumen and gums, be termed colloids. Although colloidal dispersions had certainly been studied long before this time, and the alchemists frequently worked with body fluids, which are colloidal dispersions, Graham is generally regarded as having founded the discipline of colloid science.

The test of crystal formation later turned out to be too restrictive, the distinction of crystalloids versus colloids was dropped, and the noun colloid was eventually replaced by the adjective colloidal, indicating a particular state of matter: matter for which at least one dimension falls within a specific range of distance values. The second property that distinguishes all colloidal dispersions is the extremely large area of the interface between the two phases compared with the mass of the dispersed phase. It follows that any chemical and physical phenomena that depend on the existence of an interface become very prominent in colloidal dispersions. Interface science thus underlies colloid science[2].

Now, almost 200 years later, a vast lexicon is associated with the study of colloid and interface science because, in addition to the growth of the fundamental science itself, we recognize a great diversity of occurrences and properties of colloids and interfaces in industry and indeed in everyday life. Many other scientific disciplines become involved in the study and treatment of colloidal systems, each discipline bringing elements of its own special language.

---

[1] For more historical background *see* references 1–3.

[2] Here again we encounter evidence of a living language. Following Graham's identification of a new division of chemistry, colloid chemistry, the realization of the profound importance of the interface between the phases led subsequent chemists to refer to the discipline as colloid and capillary chemistry (meaning colloid and interface chemistry). In view of the wide interdisciplinary nature it now has, I prefer the term colloid and interface science.

This book provides brief explanations for the most important terms that may be encountered in a study of the fundamental principles, experimental investigations, and industrial applications of colloid and interface science. Even this coverage represents only a personal selection of the terms that could have been included were there no constraints on the size of the book.

I have tried to include as many important terms as possible. The difficulty of keeping abreast of the colloid science vocabulary has been made worse by the tendency for the language itself to change as the science has developed, just as the meaning of the word colloid has changed. Many older terms that are either no longer in common use, or worse, that now have completely new meanings, are included as an aid to the reader of the older colloid and interface science literature and as a guide to the several meanings that many terms can have. In addition, cross-references for the more important synonyms and abbreviations are included.

Some basic knowledge of underlying fields such as physical chemistry, geology, and chemical engineering is assumed. Many named colloids and phenomena (such as Pickering emulsions) are included, but only a selection of important named equations and constants is included.

Specific literature citations are given when the sources for further information are particularly useful, unique, or difficult to find. For terms drawn from fundamental colloid and interface science, much reliance was placed on the recommendations of the IUPAC Commission on Colloid and Surface Chemistry (reference 4). In the subdisciplines several other sources were particularly helpful (references 5 –19).

Acknowledgments

This book was made possible through the support of my family, Ann Marie, Katherine, and Victoria, who gave me the time needed for the research and writing, and Conrad Ayasse, who gave me the professional freedom and encouragement needed to undertake such a project.

I thank my colleagues who invested considerable time and effort reviewing various drafts of this book and contributing helpful comments and suggestions. There are too many of these to list here, but Karin Mannhardt, Loren Hepler, Randy Mikula, and the publisher's anonymous reviewers were especially helpful. I also thank Cheryl Shanks and Janet S. Dodd of ACS Books, who provided, once again, much help and encouragement.

There are so many different, specialized references to aspects of colloidal systems in industrial practice that some important terms will inevitably have been missed. I will greatly appreciate it if readers would take the trouble to inform me of any significant errors or omissions.

Laurier L. Schramm
Petroleum Recovery Institute
100, 3512 33rd Street N.W.
Calgary, Alberta
Canada T2L 2A6

April 1993

Ablation   The reduction of particles into smaller sizes due to erosion by other particles or the surrounding fluid. May also refer to the size reduction of liquid droplets due to erosion, as in the processing of an oil sand slurry in which the oil (bitumen) is very viscous.

Absolute Viscosity   A term used to indicate viscosity measured by using a standard method, with the results traceable to fundamental units. Absolute viscosities are distinguished from relative measurements made with instruments that measure viscous drag in a fluid without known or uniform applied shear rates. *See* Viscosity, Table 4.

Absorbance   In optics, a characteristic of a substance whose light absorption is being measured. The Beer–Lambert law gives the ratio of transmitted (I) to incident (I_0) light as $\log(I/I_0) = alc$, where a is the absorptivity, l is the optical path length, and c is the concentration of species in the optical path. The logarithmic term is called the absorbance.

Absorbate   A substance that becomes absorbed into another material, or absorbent. *See* Absorption.

Absorbent   The substrate into which a substance is absorbed. *See* Absorption.

Absorption   The increase in quantity (transfer) of one material into another or of material from one phase into another phase.

Absorption may also denote the *process* of material accumulating inside another.

Acacia Gum   *See* Gum.

Acid Number   *See* Total Acid Number.

ACN   Alkane carbon number, *see* Equivalent Alkane Carbon Number.

Activated Adsorption   Chemisorption, that is, adsorption for which an activation energy barrier must be overcome, as opposed to unactivated adsorption, or physisorption, for which there is no activation energy barrier to be overcome. *See also* Chemisorption, Physisorption.

Activated Carbon   Carbonaceous material (such as coal) that has been treated, or activated, to increase the internal porosity and surface area. This treatment enhances its sorptive properties. Activated carbon is used for the removal of organic materials in water- and wastewater- treatment processes. Also termed activated charcoal.

Activated Charcoal   *See* Activated Carbon.

Activation Energy   The minimum potential energy that must be attained by a system for a reaction or process to take place at a significant rate. Catalysts usually function by providing a mechanism for a reaction that has a lower activation energy than does the uncatalyzed reaction.

Activator   Any agent that may be used in froth flotation to enhance selectively the effectiveness of collectors for certain mineral components. *See also* Froth Flotation, Depressant.

Active Site   In adsorption, the specific regions of an adsorbent onto which a substance may adsorb. In catalysis, the site responsible for a particular reaction.

Active Surfactant   The primary surfactant in a detergent formulation. *See also* Detergent.

Adatom   An adsorbed atom.

Additive Electrolyte   *See* Critical Coagulation Concentration.

Adhesion   (1) The attachment of one phase to another. *See* Work of Adhesion, Adhesive.

(2) The load causing failure of a joint, for example of a glued joint between two materials. *See* Peel Test.

Adhesion Tension   An older term that was used to refer to the wetting tension and/or the interfacial tension between solid and liquid phases. These usages have been discouraged in order to avoid confusion with the work of adhesion. *See also* reference 4.

Adhesional Wetting   The process of wetting when a surface (usually solid), previously in contact with gas, becomes wetted by liquid. This term is sometimes used to describe wetting that includes the formation of an adhesional bond between the liquid and the phase it is wetting. *See also* Wetting, Spreading Wetting, Immersional Wetting.

Adhesive   Any substance that enables or enhances mechanical adhesion, usually between solids. Example: glue.

Admicelle   *See* Hemimicelle.

Admicellar Catalysis   Catalysis occurring in the admicellar (internal) region of admicelles adsorbed onto some medium. *See also* Hemimicelle.

Admicellar Chromatography   The chromatographic separation of compounds as they pass through a medium containing media bearing admicelles. *See also* Hemimicelle.

Adsorbate   A substance that becomes adsorbed at the interface or into the interfacial layer of another material, or adsorbent. *See* Adsorption.

Adsorbent   The substrate material onto which a substance is adsorbed. *See* Adsorption.

Adsorbent Surface   *See* Adsorption Space.

Adsorption   The increase in quantity of a component at an interface or in an interfacial layer. In most usage it is positive, but it can be negative (depletion); in this sense negative adsorption is a different process from desorption. Adsorption may also denote the process of components accumulating at an interface.

Adsorption Capacity   The maximum amount of adsorbate that can be adsorbed by an adsorbent. The amount of adsorbed substance reached in a saturated solution, often where the solute is strongly adsorbed from a solution in which it has limited solubility.

Adsorption Complex   The combination of a (molecular) species that is adsorbed together with that portion of the adsorbent to which it is bound.

Adsorption Hysteresis   The phenomenon in which adsorption and desorption curves (isotherms) depart from each other.

Adsorption Isobar   *See* Adsorption Isotherm.

Adsorption Isostere   The function relating the equilibrium pressure to the equilibrium temperature for a constant value of the amount (or surface excess amount) of substance adsorbed by a specified amount of adsorbent.

Adsorption Isotherm   The mathematical or experimental relationship between the equilibrium quantity of a material adsorbed and the composition of the bulk phase, at constant temperature. The adsorption isobar is the analogous relationship for constant pressure, and the adsorption isostere is the analogous relationship for constant volume. *See also* Langmuir Isotherm, Freundlich Isotherm, Polanyi Isotherm, Gibbs Isotherm, Brunauer–Emmett–Teller Isotherm, Characteristic Isotherm.

Adsorption Site   *See* Active Site.

Adsorption Space   An interface is sometimes considered to comprise two regions, one containing a certain thickness of adsorbent and the other containing a certain thickness of the fluid phase. The former is termed the surface layer of the adsorbent (or adsorbent surface) and the latter is termed the adsorption space.

Adsorptive Material   Material that is present in one or both of the bulk phases bounding an interface and capable of becoming adsorbed.

Advancing Contact Angle   The dynamic contact angle that is measured when one phase is advancing, or increasing its area of contact, along an interface while in contact with a third, immiscible phase. It is essential to state through which phase the contact angle is measured. *See also* Contact Angle.

Advection   The transport of material solely by mass motion. In meteorology an example is the transfer of heat by horizontal motion of the air. For flow in porous media advective flow is without dispersion and results in the step appearance of chemical species at the downstream end of a control volume through which the species were flowing. In this case the Darcy velocity alone is needed to predict the "advective" time of appearance of chemical species downstream.

AEAPS   Auger electron appearance potential spectroscopy. *See* Appearance Potential Spectroscopy.

Aeolotropic   *See* Anisotropic.

Aerated Emulsion   A foam in which the liquid consists of two phases in the form of an emulsion. Also termed foam emulsion. Example: whipped cream consists of air bubbles dispersed in cream, which is an emulsion. *See also* Foam.

Aerating Agent   *See* Foaming Agent.

Aerator   (1) Any machine used for preparing foams, especially in the food industry. In batch aerators the gas is usually whipped into the liquid. In continuous aerators a mixing head is used to whip the gas into the liquid under pressure. In this case the foam expands as it leaves the machine. *See also* Oakes Mixer.

(2) In environmental applications, any means for increasing the liquid-gas interface to promote either oxygen transfer into water (e.g., to enhance microbial reactions or oxidize compounds) or to enhance the mass transfer of volatile organic materials from the liquid phase to the gas phase.

Aerogel   A special kind of xerogel where the dried-out gel retains most of the original open structure. Example: some macromolecular gels.

Aerosol   Colloidal dispersions of liquids or solids in a gas. Distinctions may be made among aerosols of liquid droplets (e.g., fog, cloud, drizzle, mist, rain, spray) and aerosols of solid particles (e.g., fume and dust). *See also* Atmospheric Aerosols of Liquid Droplets, Table 1.

AES   *See* Auger Electron Spectroscopy.

AFM   Atomic force microscopy. *See* Scanning Tunneling Microscopy, Table 9.

Agar   A water-soluble mixture of polysaccharides derived from seaweeds. Agar is considered to be composed of three types of representative structures known as neutral agarose (or agaran), pyruvic acid acetal, and sulfated galactan. The combination of the latter two types is sometimes referred to as charged agar, or agaropectin. Agar sols can be quite viscous, readily form gels, and may be used to stabilize certain suspensions, foams, and emulsions. Agar is used in many different applications including foods and medicines. *See also* reference 20, Seaweed Colloids, Hydrocolloid.

Agaran   One of the kinds of polysaccharide structure that constitutes agar. Also termed neutral agarose. *See also* Agar.

Aging   The properties of many colloidal systems may change with time in storage.

Petroleum: Aging in crude oils may refer to changes in composition due to oxidation, precipitation of components, bacterial action, or evaporation of low-boiling components.

Emulsions and foams: Aging in emulsions or foams may refer to any of aggregation, coalescence, creaming or chemical changes. Aged emulsions and foams frequently have larger droplet or bubble sizes.

Suspensions: Aging in suspensions usually refers to aggregation, that is, coagulation or flocculation. It is also used to describe the process of recrystallization, in which larger crystals grow at the expense of smaller ones, that is, Ostwald ripening.

Agglomeration   The aggregation of particles, droplets, or bubbles in a dispersion. This term is sometimes used to indicate a combination of aggregation and coalescence processes. *See* Spherical Agglomeration.

Aggregate   A group of species, usually droplets, bubbles, particles or molecules, that are held together in some way. A micelle can
be considered to be an aggregate of surfactant molecules or ions.

Aggregation   The process of forming a group of droplets, bubbles, particles, or molecules that are held together in some way.
This process is sometimes referred to interchangeably as coagulation or flocculation, although in some usage these refer to aggregation at the primary and secondary minimum, respectively. The
reverse process is termed deflocculation or peptization. *See also* Primary Minimum.

    For suspensions and emulsions, coagulation and flocculation
are frequently taken to represent two different kinds of aggregation.
In this case coagulation refers to the formation of compact aggregates, whereas flocculation refers to the formation of a loose net-
work of particles. An example can be found in montmorillonite clay
suspensions in which coagulation refers to dense aggregates produced by face-face oriented particle associations, and flocculation
refers to loose aggregates produced by edge-face- or edge-edge-ori-
ented particle associations. *See also* reference 19.

Aggregation Number   The number of surfactant molecules or
ions composing a micelle. Example: the aggregation number for
dodecyl sulfate ions in water is about 70.

Agitator Ball Mill   A machine for the comminution, or size
reduction, of minerals or other materials. Such machines crush the
input material by wet grinding in a cylindrical rotating bin containing grinding balls. Colloidal size particles can be produced with
these mills.

Algin   Any of the salt forms of alginic acid, which is a polysaccharide derived from seaweeds. Most of the salt forms are very
water soluble. Also termed alginate. Algin sols can be quite viscous,
readily form gels, and may be used to stabilize certain suspensions
and emulsions. Algin is used in many different applications including foods. *See also* reference 20, Seaweed Colloids.

Alginate   *See* Algin.

Alginic Acid   *See* Algin.

Alkane Carbon Number (ACN)   *See* Equivalent Alkane Carbon Number.

Amicron   An older particle size range distinction no longer in use. *See* Micrometer, Micron.

Amonton's Law   A description of friction that states that the coefficient of friction is given by the frictional force divided by the load normal to the direction of motion along the surfaces. *See* Friction, Lubrication.

Amott–Harvey Test   *See* Amott Test.

Amott Test   A measure of wettability based on a comparison of the amounts of water or oil imbibed into a porous medium spontaneously and by forced displacement. Amott test results are expressed as a displacement-by-oil (δ_o) ratio and a displacement-by-water ratio (δ_w). In the Amott–Harvey test a core is prepared at irreducible water saturation and then an Amott test is run. The Amott–Harvey relative displacement (wettability) index is then calculated as $\delta_w - \delta_o$, with values ranging from -1.0 for complete oil-wetting to 1.0 for complete water-wetting. *See also* reference 21, Wettability, Wettability Index.

Amott Wettability Index   *See* Amott Test.

Amphipathic   Having both lyophilic and lyophobic groups (properties) in the same molecule, as in the case of surfactants. Also referred to as being amphiphilic.

Amphiphilic   *See* Amphipathic.

Amphoteric Surfactant   A surfactant molecule for which the ionic character of the polar group depends on solution pH. For example, Lauramidopropyl betaine $C_{11}H_{23}CONH(CH_2)_3N^+(CH_3)_2CH_2COO^-$ is positively charged at low pH but is electrically neutral, having both positive and negative charges at intermediate pH. Other combinations are possible and some amphoteric surfactants are negatively charged at high pH. *See* Zwitterionic Surfactant.

Andreason Pipet   A graduated cylinder having provision for withdrawing subsamples from the bottom. Used to study sedimentation in the determination of particle sizes. *See* Stokes' Law.

Anion-Exchange Capacity   The capacity for a substrate to adsorb anionic species while simultaneously desorbing (exchanging) an equivalent charge quantity of other anionic species. Example: this property is sometimes used to characterize clay minerals that often have very large cation-exchange capacities but may also have significant anion-exchange capacities. *See also* Ion Exchange.

Anionic Surfactant   A surfactant molecule that can dissociate to yield a surfactant ion whose polar group is negatively charged. Example: sodium dodecyl sulfate, $CH_3(CH_2)_{11}SO_4^-Na^+$.

Anisokinetic Sampling   *See* Isokinetic Sampling.

Anisotropic   A material that exhibits a physical property, such as light transmission, differently in different directions. Sometimes termed aeolotropic.

Anneal   The process of heating a solid material to a temperature close to, but lower than, its melting point to reduce internal stresses and strengthen the material.

Anode   An electrode at which a net positive current flows. The predominant chemical reaction here is oxidation.

Anomalous Water   *See* Polywater.

Antagonistic Electrolyte   *See* Critical Coagulation Concentration.

Anti-Bubbles   A dispersion of liquid-in-gas-in-liquid wherein a droplet of liquid is surrounded by a thin layer of gas that in turn is surrounded by bulk liquid. Example: in an air-aqueous surfactant solution system this would be designated as water-in-air-in-water, or W/A/W, in fluid film terminology. A liquid–liquid analogy can be drawn with the structures of multiple emulsions. *See also* reference 22, Fluid Film.

Antifoaming Agent   Any substance that acts to reduce the stability of a foam; it may also act to prevent foam formation. Terms such as antifoamer or foam inhibitor are used to specify the prevention of foaming, and terms such as defoamer or foam breaker are used to specify the reduction or elimination of foam stability. Example: poly(dimethylsiloxane)s, $(CH_3)_3SiO[(CH_3)_2SiO]_xR$, where R represents any of a number of organic functional groups. Antifoamers may act by any of a number of mechanisms.

Antonow's Rule   An empirical rule for the estimation of interfacial tension between two liquids as the difference between the surface tensions of each liquid. Even for pure liquids this rule is seldom very accurate. *See* Table 8.

A/O/W   An abbreviation for a fluid film of oil between air and water. Usually designated W/O/A. *See* Fluid Film.

APD   Azimuthal photoelectron diffraction. *See* Photoelectron Diffraction.

Aphrons   *See* Microgas Emulsions.

API Gravity   A measure of the relative density (specific gravity) of petroleum liquids. The API gravity, in degrees, is given by $°API = (141.5/\text{relative density}) - 131.5$, where the relative density at temperature t (°C) = (density at t)/(density of water at 15.6 °C).

Apolar   Description applied to materials or surfaces that have no polar nature.

Apparent Viscosity   Viscosity determined for a non-Newtonian fluid without reference to a particular shear rate for which it applies. Such viscosities are usually determined by a method strictly applicable to Newtonian fluids only. *See* Table 4.

Appearance Potential Spectroscopy (APS)   A technique related to photoelectron spectroscopy and also used for the determination of surface composition. The surface is scanned with an electron beam of varying energy, which causes the ejection of inner electrons from the surface atoms. The intensities of the beams of ejected electrons are determined (X-ray or Auger electrons). The terms Auger electron appearance potential spectroscopy (AEAPS) and soft X-ray appearance potential spectroscopy (SXAPS) are used to distinguish modes in which Auger electrons or photons, respectively, are emitted. *See also* Table 9.

APS   *See* Appearance Potential Spectroscopy.

Areal Elasticity   *See* Film Elasticity.

Asphalt   A naturally occurring hydrocarbon that is a solid at reservoir temperatures. An asphalt residue may also be prepared from heavy (asphaltic) crude oils or bitumen, from which lower boiling fractions have been removed.

Asphaltene   A high-molecular-mass, polyaromatic component of some crude oils that also has high sulfur, nitrogen, oxygen, and metal contents. In practical work asphaltenes are usually defined operationally by using a standardized separation scheme. One such scheme defines asphaltenes as those components of a crude oil or bitumen that are soluble in toluene but insoluble in *n*-pentane.

Association Colloid   A dispersion of colloidal-sized aggregates of small molecules; it is lyophilic. Example: micelles of surfactant molecules or ions in water.

Atactic Polymer   In polymers having monomer units of the form (CH_2CHR), the hydrogen atoms and R groups may tend to align differently with respect to an imaginary plane containing the carbon atom chain. In an atactic polymer the orientation is random; in a syndiotactic polymer the R groups alternate from side to side; in an isotactic polymer all the R groups lie on one side and all the hydrogen atoms lie on the other.

Atmospheric Aerosols of Liquid Droplets   Atmospheric colloidal dispersions of liquids in gas. Distinctions are made among fog, cloud, drizzle, and rain, depending upon droplet sizes and based on whether the droplets would be large enough to fall to the earth's surface before completely evaporating. Droplets small enough to evaporate before reaching the ground fall into the fog and cloud ranges, less than about 100 μm. Droplets greater than about 100 μm are large enough to reach the ground before evaporating and fall into the rain category. Droplets with diameters of about 100 μm are termed drizzle and correspond to a stage in the development of rain through the coalescence of cloud droplets. *See* Table 1.

Atomic Force Microscopy (AFM)   *See* Scanning Tunneling Microscopy, Table 9.

Atterberg Limits   A group of (originally) seven limits of soil consistency, or relative ease with which material can be deformed or made to flow. The only Atterberg limits that are still in common use are the liquid limit, plastic limit, and plasticity number. *See* references 23 and 24.

Attractive Potential Energy   *See* Gibbs Energy of Attraction.

Auger Electron Appearance Potential Spectroscopy (AEAPS)   *See* Appearance Potential Spectroscopy.

Auger Electron Spectroscopy (AES)   A technique used for the determination of surface composition by scanning the surface with an electron beam. The beam ionizes surface atoms by ejecting inner-shell electrons. Electron transfer from outer electron shells will result in the emission of energy, either as characteristic X-rays, or in the ejection of a second outer-shell electron (Auger electron). Auger electrons have energies characteristic of the atoms from which they were ejected. *See also* Table 9.

Autophobicity   *See* Spreading Coefficient.

A/W/A   An abbreviation for a fluid film of water in air. *See* Fluid Film.

A/W/O   An abbreviation for a fluid film of water between air and oil phases. Also termed pseudoemulsion film. Usually designated O/W/A. *See* Fluid Film.

Azimuthal Photoelectron Diffraction (APD)   *See* Photoelectron Diffraction.

Backscattering   *See* Light Scattering.

Bacteria   Small single-celled microorganisms, typically between 200 and 2000 nm in diameter or length. They can form stable colloidal dispersions in water because they tend to have a net negative surface charge.

Bancroft's Rule   An empirical generalization that predicts that the continuous phase in an emulsion will be the phase in which the emulsifying agent is most soluble. An extension for solid particles acting as emulsifying agents predicts that the continuous phase will be the phase that preferentially wets the solid particles. *See also* Hydrophile–Lipophile Balance.

Basic Sediment and Water   That portion of solids and aqueous solution in an emulsion that separates out on standing or is separated by centrifuging in a standardized test method. Basic sediment may contain emulsified oil as well. Also referred to as BS&W, BSW, Bottom Settlings and Water, and Bottom Solids and Water.

Batch Treating   In oil production or processing, the process in which emulsion is collected in a tank and then broken in a batch. This process is as opposed to continuous or flow-line treating of emulsions.

Bed Knives   The stationary cutting blades in a cutting mill machine for comminution.

Beer–Lambert Law   *See* Absorbance.

Beneficiation   In mineral processing, any process that results in a product having an improved desired mineral content. Example: froth flotation.

BET Isotherm   *See* Brunauer–Emmett–Teller Isotherm.

Bicontinuous System   A two-phase system in which both phases are continuous phases. For example, a possible structure for middle-phase microemulsions is one in which both oil and water phases are continuous throughout the microemulsion phase. An analogy can be drawn from the structure of porous and permeable rock in which both the mineral phase and the pore or throat channels can be continuous at the same time. *See also* Middle-Phase Microemulsion.

Bilayer   *See* Bimolecular Film.

Biliquid Foam   A concentrated emulsion of one liquid dispersed in another liquid.

Bimolecular Film   A membrane that separates two aqueous phases and is composed of two layers of polar organic molecules, such as surfactants or lipids, that are oriented with their hydrocarbon groups in the two molecular layers towards each other and the polar groups facing the respective aqueous phases. *See also* Vesicle, Black Lipid Membrane.

Bingham Plastic Fluid   *See* Plastic Fluid.

Biocolloidal Dispersion   A colloidal dispersion in which the dispersed phase is of biological origin. Example: a dispersion of lipid particles.

Bipolar   A substance having electron-donor as well as electron-acceptor properties. This feature has an influence on surface tension. Bipolar is not the same as dipolar. *See also* Dipole.

Birefringent   A material that has different refractive indices in different directions. Example: liquid crystals.

Bitumen   A naturally occurring viscous hydrocarbon having a viscosity greater than 10,000 mPa·s at ambient deposit temperature, and a density greater than 1000 kg/m^3 at 15.6 °C. *See* references 25–27. In addition to high-molecular-mass hydrocarbons, bitumen contains appreciable quantities of sulfur, nitrogen, oxygen, and heavy metals.

Black Film   Fluid films yield interference colors in reflected white light that are characteristic of their thickness. At a thickness of about 0.1 μm the films appear white and are termed silver films. At reduced thicknesses they first become grey and then black (black films). Among thin equilibrium (black) films, one may distinguish those that correspond to a primary minimum in interaction energy, typically at thicknesses of about 5 nm (Newton black films) from those that correspond to a secondary minimum, typically at thicknesses of about 30 nm (common black films).

Black Lipid Membrane   A bimolecular film in which the molecules composing the membrane film are lipid molecules. The term "black" refers to the fact that these films appear black when illuminated (no apparent interference colors). *See* Bimolecular Film, Black Film.

Blender Test   An empirical test in which an amount of potential foaming agent is added into a blender containing a specified volume of liquid to be foamed. After blending at a specified speed and for some specified time, the blending is halted and the extent (volume) of foam produced is measured both immediately and after a period of time of quiescent standing. There are many variations of this test. *See also* Bottle Test.

BLM   *See* Black Lipid Membrane.

Block Copolymer   A polymer composed of two kinds of monomers in which the repeating unit comprises a chain, or block, of several units of each monomer type.

Boltzmann Equation   A fundamental equation giving the local concentrations of ions in terms of the local electric potential, in an electric double layer. *See also* Poisson–Boltzmann Equation.

Böttcher Equation   For predicting the relative permittivity of dispersions. *See* Table 7.

Bottle Test   Emulsions: An empirical test in which varying amounts of a potential demulsifier or coagulant are added into a series of tubes or bottles containing subsamples of an emulsion or other dispersion that is to be broken or coagulated. After some specified time the extent of phase separation and appearance of the interface separating the phases are noted. There are many variations of this test. For emulsions, in addition to the demulsifier, a diluent may be added to reduce viscosity. In the centrifuge test, centrifugal force may be added to speed up the phase separation. Other synonyms include jar test, beaker test.

Foams: An empirical test in which an amount of potential foaming agent (or even defoaming agent) is added into a bottle containing a specified volume of liquid to be foamed. After shaking the bottle in a specific manner and for some specified time, the shaking is halted and the extent (volume) of foam produced is measured both immediately and after a period of time of quiescent standing. There are many variations of this test. *See also* Blender Test.

Water treatment: A standard test method in which either the coagulant dosage is varied or the solution pH is varied for a given coagulant dosage, to optimize the coagulation of solids. Frequently termed jar test.

Bottom Settlings and Water   *See* Basic Sediment and Water.

Boundary Lubrication   *See* Lubrication.

Boussinesq Number   A measure of the ratio of interfacial and bulk viscous effects in a thinning foam film: $B_o = (\eta^s + \kappa^s)/(\eta R_f)$ where η^s is the surface shear viscosity, κ^s is the surface dilational viscosity, η is the bulk liquid viscosity, and R_f is the thin film radius.

Breaking   The process in which an emulsion or foam separates. Usually coalescence causes the separation of a macrophase, and eventually the formerly dispersed phase becomes a continuous phase, separate from the original continuous phase.

Bredig Arc Method   A method in which metal particles are dispersed by passing an electric current between two wires (forming the arc) immersed in a liquid.

Brighteners   *See* Optical Brighteners.

Brightening Agents   *See* Optical Brighteners.

Bright-Field Illumination   A kind of illumination for micro-scopy, in which the illumination of a specimen is arranged so that transmitted light remains in the optical path of the microscope and is used to form the magnified image. This is different from the arrangement in Dark-Field Illumination.

Brownian Motion   Random fluctuations in the density of mole-cules at any location in a liquid, due to thermal energy, cause other molecules and small dispersed particles to move along random pathways. The random particle motions are termed Brownian motion and are most noticeable for particles smaller than a few micrometers in diameter.

Bruggeman Equation   An equation for predicting the conduc-tivities or relative permittivities of dispersions. *See* Tables 6 and 7.

Brunauer–Emmett–Teller Isotherm (BET Isotherm)   An adsorption isotherm equation that accounts for the possibility of multilayer adsorption and different enthalpy of adsorption between the first and subsequent layers. Five "types" of adsorption isotherm are usually distinguished. These are denoted by roman numerals and refer to different characteristic shapes. *See* Adsorption Isotherm.

BS&W   *See* Basic Sediment and Water.

Bubble Snap-Off   *See* Snap-Off.

Builder   A chemical compound added into detergent formula-tions to aid oil emulsification (by raising pH) and to complex and solubilize hardness ions. Example: sodium tripolyphosphate.

Bulk Foam   Any foam for which the length scale of the confining space is greater than the length scale of the foam bubbles. The con-verse case categorizes some foams in porous media, distinguished by the term "lamellar foam". *See also* Foam, Foam Texture.

Cabannes Factor A factor used in light-scattering analysis to correct for particle anisotropy. *See* Depolarization.

Canal Viscometer An instrument used to measure interfacial viscosity by measuring the flow rate of a surface or interfacial layer through a narrow channel or canal. The analysis is essentially a two-dimensional analog of the capillary viscosity method for fluids.

Capacitance of the Electric Double Layer The integral capacitance of the electric double layer (per unit area) is the charge density at the outer Helmholtz plane divided by the electric potential at the outer Helmholtz plane. The differential capacitance of the electric double layer (per unit area) is the partial derivative of the charge density with respect to the potential at the outer Helmholtz plane.

Capillarity A general term referring either to the general subject of, or to the various phenomena attributable to the forces of surface or interfacial tension. The Young–Laplace equation is sometimes referred to as the equation of capillarity.

Capillary A tube having a very small internal diameter. Originally the term referred to cylindrical tubes whose internal diameters were of similar dimension to hairs.

Capillary Condensation The process by which multilayer adsorption from a vapor into a porous medium proceeds to the point at which pore spaces become filled with condensed liquid from the vapor. The dimensions of the pore must be large enough that the concept of a separating meniscus retains a physical meaning.

Capillary Constant   For two phases in contact the capillary constant, a, is given by $a^2 = 2\gamma/\Delta\rho g$; that is, the square of the capillary constant equals the ratio of twice the surface (or interfacial) tension to the product of gravitational constant and the density difference between the phases. This dimensionless number is used in considerations of capillarity, such as in capillary rise.

Capillary Electrometer   An instrument used to determine electrocapillary curves. A column of mercury is attached to an electrochemical cell that is used to apply an electric potential to the mercury–aqueous solution interface. The interfacial tensions corresponding to different states of applied electric potential were originally determined by capillary rise, but subsequent capillary electrometers have used other interfacial tension methods. *See* Electrocapillarity.

Capillary Flow   Liquid flow in response to a difference in pressures across curved interfaces. *See also* Capillary Pressure.

Capillary Forces   The interfacial forces acting among oil, water, and solid in a capillary or in a porous medium. These determine the pressure difference (capillary pressure) across an oil–water interface in the capillary or in a pore. Capillary forces are largely responsible for oil entrapment under typical petroleum reservoir conditions.

Capillary Number (N_c)   A dimensionless ratio of viscous to capillary forces. One form gives N_c as Darcy velocity times viscosity of displacing phase divided by interfacial tension. It is used to provide a measure of the magnitude of forces that trap residual oil in a porous medium.

Capillary Pressure   The pressure difference across an interface between two phases. When the interface is contained in a capillary, it is sometimes referred to as the suction pressure.

In petroleum reservoirs it is the local pressure difference across the oil–water interface in a pore contained in a porous medium. One of the liquids usually preferentially wets the solid, and therefore the capillary pressure is normally taken as the pressure in the nonwetting fluid minus that in the wetting fluid.

Capillary Ripples   Surface or interfacial waves caused by perturbations of an interface. Where the perturbations are caused by mechanical means (e.g., barrier motion) the transverse waves are known as capillary ripples or Laplace waves, and the longitudinal waves are known as Marangoni waves. The characteristics of these waves depend on the surface tension and the surface elasticity. This feature forms the basis for the capillary wave method of determining surface or interfacial tension.

Capillary Rise   The tendency, and process, for a liquid to rise in a capillary. Example: water rises in a partially immersed glass capillary. Negative capillary rise occurs when the liquid level in the capillary falls below the level of bulk liquid, as when a glass capillary is partially immersed in mercury. Capillary rise forms the basis for a method of determination of surface or interfacial tension.

Capillary Viscometer   An instrument used for the measurement of viscosity in which the rate of flow through a capillary under constant applied pressure difference is determined. This method is most suited to the determination of Newtonian viscosities. There are various designs, among which are the Ostwald and Ubbelohde types.

Capillary Wave Method   *See* Capillary Ripples.

Capillary Waves   *See* Capillary Ripples.

Carbon Black   Carbon particles of very small size, hence large specific surface area, capable of acting as an effective adsorbent for some substances.

Carrageenan   A water-soluble mixture of sulfated linear polysaccharides derived from seaweeds such as Irish Moss. Carrageenan is considered to have a number of different structural types that are designated by different Greek letter prefixes, for example κ-carrageenan. Carageenan sols can be quite viscous and readily form gels, and they may be used to stabilize certain suspensions, foams, and emulsions. Carrageenan is used in many different applications including foods. *See also* reference 20, Seaweed Colloids, Hydrocolloid.

Catalyst   A substance that increases the rate or yield of a reaction. Heterogeneous catalysis refers to the situation in which the catalytic reactions occur at a surface or interface between two phases. In practice heterogeneous catalysts tend to have high specific surface areas. Homogeneous catalysis refers to the situation in which the reaction(s) take place within a single phase.

Catalyst Poison   *See* Poison.

Catalyst Promoter   *See* Promoter.

Cataphoresis   *See* Electrophoresis.

Cathode   An electrode at which a net negative current flows. The predominant chemical reaction here is reduction.

Cation-Exchange Capacity   The capacity for a substrate to adsorb hydrated cationic species while simultaneously desorbing (exchanging) an equivalent charge quantity of other cationic species. Example: this property is used to characterize clay minerals that may have very large cation-exchange capacities and also significant anion-exchange capacities. *See also* Ion Exchange.

Cationic Surfactant   A surfactant molecule that can dissociate to yield a surfactant ion whose polar group is positively charged. Example: cetyltrimethylammonium bromide, $CH_3(CH_2)_{15}N^+(CH_3)_3Br^-$.

CCC   *See* Critical Coagulation Concentration.

CCT   *See* Critical Coagulation Temperature.

Cell Membrane   Thin films composed of lipids and proteins that cover the surfaces of cells. Also termed plasma membranes or plasmalemma.

CELS   *See* Characteristic Energy-Loss Spectroscopy.

Centrifugal Separator   *See* Separator.

Centrifuge   An apparatus in which an applied centrifugal force is used to achieve a phase separation by sedimentation or creaming. For centrifuges operating at very high relative centrifugal forces (so-called *g*-forces) the terms supercentrifuge (ca. tens of thousands RCF or *g*s) or ultracentrifuge (ca. hundreds of thousands RCF or *g*s) are used. *See also* Relative Centrifugal Force.

Centrifuge Test   *See* Bottle Test.

CFC   Critical flocculation concentration. *See* Critical Coagulation Concentration.

CFT   Critical flocculation temperature. *See* Critical Coagulation Temperature.

Characteristic Adsorption Curve   *See* Characteristic Isotherm.

Characteristic Energy-Loss Spectroscopy (CELS)   A technique for studying surface composition and surface energy states. Inelastically scattered electrons, having lower energy than the incident beam, are used to form the image pattern, and the characteristic energy losses of the scattered electrons are determined. Similar techniques include electron loss spectroscopy (ELS), which is also termed electron energy loss spectroscopy (EELS), electron impact spectroscopy (EIS), and high-resolution electron energy loss spectroscopy (HREELS).

Characteristic Isotherm   An adsorption isotherm involving multilayer adsorption in which the equilibrium quantity of a material adsorbed is essentially related to the composition of the bulk phase, at constant temperature, by a single relationship for a given adsorbate independently of the nature of the adsorbent. Also termed the characteristic adsorption curve. *See also* Adsorption Isotherm.

Charged Agar   A combination of two of the types of polysaccharide structure that constitute agar, pyruvic acid acetal, and sulfated galactan. Charged agar is also termed agaropectin. *See also* Agar.

Charge Density   In colloidal systems, the quantity of charge at an interface, expressed per unit area.

Charge of the Micelle   *See* Micellar Charge.

Charge Reversal   The process wherein a charged substance is caused to take on a new charge of the opposite sign. Such a change can be brought about by any of oxidation, reduction, dissociation, ion exchange, or adsorption. Example: the adsorption of cationic polymer molecules onto negatively charged clay particles can exceed the requirements for charge neutralization and thus cause charge reversal.

Chemical Adsorption   *See* Chemisorption.

Chemisorption (Chemical adsorption)　　The adsorption forces are of the same kind as those involved in the formation of chemical bonds. The term chemisorption is used to distinguish chemical adsorption from physical adsorption, or physisorption, in which the forces involved are of the London-van der Waals type. Some guidelines for distinguishing between chemisorption and physisorption are given by IUPAC in reference 4.

Chi Potential　　*See* Jump Potential.

Chitin　　A mucopolysaccharide, soluble in organic liquids, derived from various plants, fungi, and some marine animals, such as shellfish. Derivatized (deacetylated) chitin is referred to as chitosan. Chitin and chitosan solutions can be quite viscous, depending on the solvent, usually an organic liquid or an acid solution. They are used as coagulating agents in food industry applications. *See also* reference 20, Marine Colloids.

Chitosan　　Derivatized (deacetylated) chitin. *See also* Chitin.

Chocolate Mousse Emulsion　　A name frequently used to refer to the W/O emulsions of high water content that are formed when crude oils are spilled on the oceans. The name reflects the color and very viscous consistency of these emulsions. It has also been applied to other petroleum emulsions of similar appearance.

Chromatography　　A process or procedure in which flow through a permeable porous medium or through a capillary causes components in a mixture to become separated as a result of their different affinities for the mobile and stationary phases. Several kinds of liquid chromatography are capable of differentiating large molecules on the basis of size: In gel filtration (gel permeation) chromatography the stationary phase is a polymer gel or porous bead packing capable of sorbing smaller size molecules while larger size molecules pass through. The term can also refer to the separation of subsurface contaminant plumes in the environment. *See also* Hydrodynamic Chromatography.

Classifier　　A machine used to separate particles of specified size ranges. Wet classifiers include settling tanks, centrifuges, hydrocyclones, and vibrating screens. Dry classifiers, also termed air classifiers, employ gravity or centrifugal settling in gas streams. *See also* reference 28.

Classifier Mill   A kind of mechanical impact mill or jet mill for size reduction (comminution) that also incorporates a particle classifier.

Clausius–Mossotti Factor   An equation reflecting differences in permittivity between a colloidal species (ε_c) and the surrounding medium (ε_m) as $K = (\varepsilon_c - \varepsilon_m)/(\varepsilon_c + 2\varepsilon_m)$. K, the Clausius–Mossotti factor, is used in dielectrophoresis. If $K > 0$, the motion of a dipolar colloidal species is toward the most intense region of an imposed, non-homogeneous electric field. If $K < 0$ the species moves toward the least intense region, and for $K = 0$ there is no dielectrophoretic motion. See also Dielectrophoresis, Levitator.

Clays   (1) The term clay minerals refers to the aluminosilicate minerals having two- or three-layer crystal structure. These minerals typically exhibit high specific surface area, significant surface charge density (cation-exchange capacity), and low hydraulic conductivity. Examples: montmorillonite, kaolinite, illite.

(2) The term clays is sometimes used to distinguish particles having sizes of less than about 2 to 4 µm, depending upon the size classification system used. In this sense the term includes any suitably fine-grained solids, including nonclay minerals.

Cloud   *See* Atmospheric Aerosols of Liquid Droplets, Table 1.

Cloud Point   The transition temperature above which a nonionic surfactant loses some of its water solubility and becomes ineffective as a surfactant. The originally transparent surfactant solution becomes cloudy because of the separation of a surfactant-rich phase. Cloud points are typically reported on the basis of tests for a specified surfactant concentration such as 1 mass%. *See also* Coacervation.

CMC   *See* Critical Micelle Concentration.

Coacervate   *See* Coacervation.

Coacervation   When a lyophilic colloid loses stability, a separation into two liquid phases may occur. This process is termed coacervation. The phase that is more concentrated in the colloid is the coacervate, and the other phase is the equilibrium solution. *See also* Cloud Point.

Coactive Surfactant   The secondary surfactant(s) in a detergent formulation. *See also* Detergent.

Coadsorption   The adsorption of more than one species simultaneously.

Coagulation   *See* Aggregation.

Coagulum   The dense aggregates formed in coagulation are referred to, after separation, as coagulum. *See* Aggregation.

Coalescence   The merging of two or more dispersed species into a single one. Coalescence reduces the total number of dispersed species and also the total interfacial area between phases. In emulsions and foams coalescence can lead to the separation of a macrophase, in which case the emulsion or foam is said to break. The coalescence of solid particles is termed sintering.

Coarse Sand   *See* Sand, Table 3.

Coefficient of Friction   *See* Friction.

Cohesive Energy Density   *See* Solubility Parameter.

Co-ions   In systems containing large ionic species (colloidal ions, membrane surfaces, etc.), co-ions are those that, compared to the large ions, have low molecular mass and the same charge sign. For example, in a suspension of negatively charged clay particles containing dissolved sodium chloride, the chloride ions are co-ions and the sodium ions are counterions. *See also* Counterions.

Collapse Pressure   The film pressure required to cause a surface or interfacial monomolecular film to compress to an area that will no longer support a monolayer of adsorbed species; thus it will distort and collapse.

Collector   A surfactant used in froth flotation to adsorb onto solid particles, make them hydrophobic, and thus facilitate their attachment to gas bubbles. *See also* Froth Flotation.

Colligative Properties   Properties of matter that depend upon the number of species rather than upon their mass or activity.

Colloid   In the early 1800s Thomas Graham studied the diffusion, osmotic pressure, and dialysis properties of a number of substances, including a variety of solutes dissolved in water (*see* references 1–3). Some substances diffused quite quickly through parchment paper and animal membranes and formed crystals when dried. Other substances diffused only very slowly if at all through the parchment or membranes and apparently did not form crystals when dried. Graham proposed that the former group of substances, which included simple salts, be termed crystalloids and the latter group, which included albumen and gums, be termed colloids. The test of crystal formation later turned out to be too restrictive, the distinction of crystalloids versus colloids was dropped, and the noun colloid was eventually replaced by the adjective Colloidal. *See also* Colloidal Dispersion.

Colloidal   A state of subdivision in which the particles, droplets, or bubbles dispersed in another phase have at least one dimension between about 1 and 1000 nm. *See also* Colloidal Dispersion.

Colloidal Dispersion   A system in which colloidal species are dispersed in a continuous phase of different composition or state. *See* Table 2.

Colloidal Electrolyte   An electrolyte that dissociates to yield ions at least one of which is of colloidal or near-colloidal size. Example: ionic surfactant micelles.

Colloidal Gas Aphrons   *See* Microgas Emulsions.

Colloidal Processing   In ceramics, a variation of slip-casting in which a stabilized colloidal dispersion of particles is poured into a mold for sintering.

Colloid Anion   A colloidal species having a net negative electric charge.

Colloid Cation   A colloidal species having a net positive electric charge.

Colloid Mill   A high-shear mixing device used to prepare colloidal dispersions of particles or droplets by size reduction (comminution). Also termed dispersion mill.

Colloid Osmotic Pressure   When a colloidal system is separated from its equilibrium liquid by a semipermeable membrane, not permeable to the colloidal species, the colloid osmotic pressure is the pressure difference required to prevent transfer of the dissolved, noncolloidal species. Also referred to as the Donnan pressure. The reduced osmotic pressure is the colloid osmotic pressure divided by the concentration of the colloidal species. *See also* Osmotic Pressure.

Colloid Stability   In colloid science the term colloid stability means that a specified process that causes the colloid to become a macrophase, such as aggregation, does not proceed at a significant rate. Colloid stability is different from thermodynamic stability (*see* reference 4). The term colloid stability must be used with reference to a specific and clearly defined process, for example, a colloidally metastable emulsion may signify a system in which the droplets do not participate in aggregation, coalescence, or creaming at a significant rate. *See also* Kinetic Stability, Thermodynamic Stability.

Colloid Titration   A method for the determination of charge, and the zero point of charge, of colloidal species. The colloid is subjected to a potentiometric titration with acid or base to determine the amounts of acid or base needed to establish equilibrium with various pH values. By titrating the colloid in different, known concentrations of indifferent electrolyte, the point of zero charge can be determined as the pH for which all the isotherms intersect. *See also* Point of Zero Charge.

Comminution   The reduction of particles, or other dispersed species, into smaller sizes. Examples of comminution machines include agitator ball mills, colloid mills, cutting mills, disk mills, jet mills, mechanical impact mills, ring-roller mills, and roll crushers. *See also* reference 28, Ablation.

Common Black Film   *See* Black Film.

Compaction   *See* Subsidence.

Complex Coacervation   The process of coacervation when caused by the interaction of oppositely charged colloids.

Composite   A material that contains particles of a second substance introduced to increase material strength.

Composite Isotherm   A (no longer recommended) term referring to the Surface Excess Isotherm.

Compressibility   (1) The ratio of relative volume change to applied compressional stress.

(2) For film compressibility, *see* Film Elasticity.

Compressional Modulus   *See* Film Elasticity.

Concentric Cylinder Rheometer   *See* Rheometer.

Condensate   Any light hydrocarbon liquid mixture obtained from the condensation of hydrocarbon gases. Condensate typically contains mostly propane, butane, and pentane.

Condensation Methods   The class of methods used for preparing colloidal dispersions in which either precipitation from solution or chemical reaction is used to create colloidal species. The colloidal species are built up by deposition on nuclei that may be of the same or different chemical species. If the nuclei are of the same chemical species, the process is referred to as homogeneous nucleation; if the nuclei are of different chemical species, the process is referred to as heterogeneous nucleation. *See also* Dispersion Methods.

Condensed State   In adsorption, the state of a monolayer when it can be considered to behave like a layer of liquid (close-packed molecules).

Conductivity of a Dispersion   *See* Table 6.

Cone-Cone Rheometer   *See* Rheometer.

Cone-Plate Rheometer   *See* Rheometer.

Confocal Microscopy   A microscopic technique used to produce three-dimensional images of specimens that actually have considerable depth. A series of shallow depth-of-field image slices through a thick specimen are obtained. The three-dimensional image is then obtained by reconstruction so that no out-of-focus elements contribute to the final image.

Consistency   An empirical or qualitative term referring to the relative ease with which a material can be deformed or made to flow. It is a reflection of the cohesive and adhesive forces in a mixture or dispersion. *See also* Atterberg Limits.

Contact Angle   When two immiscible fluids (e.g., liquid–gas or oil–water) are both in contact with a solid, the angle formed between the solid surface and the tangent to the fluid–fluid interface intersecting the three-phase contact point is termed the contact angle. It is essential to state through which phase the contact angle is measured. By convention, if one of the fluids is water then the contact angle is measured through the water phase; otherwise, the contact angle is usually measured through the most dense phase. Distinctions may be made among advancing, receding, or equilibrium contact angles. Contact angles are important in areas such as liquid wetting, imbibition, and drainage.

Contact-Angle Hysteresis   A phenomenon manifested by differing values of advancing and receding contact angles in the same three-phase contact system. Both may differ from the equilibrium contact angle. *See also* Contact Angle.

Continuous Phase   In a colloidal dispersion, the phase in which another phase of particles, droplets, or bubbles is dispersed. Sometimes referred to as the external phase. Continuous phase is the opposite of dispersed phase. *See also* Dispersed Phase.

Copolymer   A polymer composed of more than one kind of monomer. *See also* Block Copolymer.

Cosorption Lines   Contours of equal surface activity, as measured by the Gibbs surface excess concentrations, plotted on phase diagrams. See reference 32, p 131.

Cosurfactant   A surfactant that may be added to a system to enhance the effectiveness of another surfactant. The term cosurfactant has also been improperly used to describe non-surface- active species that enhance a surfactant's effectiveness, such as an alcohol or a builder.

Couette Flow   The flow of liquid in the annulus between two concentric cylinders that rotate at different speeds. In the Couette rheometer one cylinder rotates, and torque is measured at the other. *See also* Rheometer.

Couette Rheometer   *See* Couette Flow.

Coulter Counter Technique   A particle- or droplet-sizing technique in which the flow of dispersed species in a capillary, between charged electrodes, causes changes in conductivity that are interpreted in terms of the sizes of the species. Coulter is the brand name for the automated counter. *See also* Sensing Zone Technique.

Counterions   In systems containing large ionic species (colloidal ions, membrane surfaces, etc.), counterions are those that, compared to the large ions, have low molecular mass and opposite charge sign. For example, clay particles are usually negatively charged and are naturally associated with exchangeable counterions such as sodium and calcium. In the early literature the term Gegenion was used to mean counterion. *See also* Co-ions.

Cream   The process of creaming in a dilute emulsion usually produces a discernible, more concentrated emulsion termed cream and having a volume termed the cream volume.

Creaming   The process of emulsion droplets floating upwards under gravity or in a centrifugal field to form a concentrated emulsion (cream) quite distinct from the underlying dilute emulsion. This is not the same as the breaking of an emulsion. *See also* Sedimentation.

Cream Volume   *See* Cream.

Creeping Flow   Gradual deformation under an applied stress. *See also* Viscoelastometer.

Critical Coagulation Concentration   (CCC) The electrolyte concentration that marks the onset of coagulation of dispersed species. The CCC is very system-specific, although the variation in CCC with electrolyte composition has been empirically generalized. *See also* Schulze–Hardy Rule. If the CCC in a binary electrolyte of salts A and B occurs at concentrations c_A and c_B compared to the pure component CCCs of c°_A and c°_B, then three types of additive effects may be distinguished (*see* reference 4) as follows. The electrolytes are

$$\text{additive if } (c_A/c^{\circ}_A) + (c_B/c^{\circ}_B) = 1$$
$$\text{antagonistic if } (c_A/c^{\circ}_A) + (c_B/c^{\circ}_B) > 1$$
$$\text{synergistic if } (c_A/c^{\circ}_A) + (c_B/c^{\circ}_B) < 1$$

Critical Coagulation Temperature   (CCT) The minimum temperature to which a dispersion must be raised in order to induce coagulation. See also Critical Coagulation Concentration.

Critical Film Thickness   A fluid film may thin to a narrow range of film thicknesses within which it either becomes metastable to thickness changes (equilibrium film) or else ruptures. Persistent foams comprise fluid films at their critical film thickness.

Critical Flocculation Concentration   *See* Critical Coagulation
Concentration.

Critical Flocculation Temperature   *See* Critical Coagulation
Temperature.

Critical Micelle Concentration   (c.m.c. or CMC) The surfactant
concentration above which molecular aggregates, termed micelles,
begin to form. In practice a narrow range of surfactant concentrations represents the transition from a solution in which only single,
unassociated surfactant molecules (monomers) are present to a
solution containing micelles. Useful tabulations are given in references 15 and 16. *See also* Micelle.

Critical Surface Tension of Wetting   The minimum, or transition, surface tension of a liquid for which it will no longer exhibit
complete wetting of a solid. This value is usually taken to be characteristic of a given solid and is sometimes used as an estimate of the
solid's surface tension. For a given solid it is typically determined by
plotting the cosine of contact angles between the solid of interest
and a series of liquids versus the surface tensions of those liquids (a
Zisman plot). The surface tension extrapolated to zero contact angle
is the critical surface tension of wetting of the solid. *See also*
Hydrophobic Index.

Critical Temperature   In adsorption, the transition temperature
at which a monolayer no longer exhibits the properties of a condensed state.

Critical Thickness   *See* Critical Film Thickness.

Critical Wetting Surface Tension   *See* Critical Surface Tension
of Wetting.

Crude Oil   A naturally occurring hydrocarbon produced from an
underground reservoir. In the petroleum field, distinctions drawn
among light, heavy, extra-heavy, and bituminous crude oils are
made. *See also* references 25–27, Oil, Light Crude Oil, Heavy Crude
Oil, Extra Heavy Crude Oil, Bitumen, Asphalt.

Cryogenic SEM   *See* Freeze Fracture Method, Scanning Electron Microscopy.

Cryogenic TEM   *See* Freeze Fracture Method, Transmission Electron Microscopy.

Crystalloid   *See* Colloid.

Cuff-Layer Emulsion   *See* Interface Emulsion.

Curd-Fibers   *See* Soap Curd.

Curie Point   The temperature above which a ferromagnetic material becomes paramagnetic.

Cutting Mill   A machine for the comminution, or size reduction, of materials. Such machines use a rotating shaft on which is mounted a series of cutting knives that interleave with a series of separately mounted stationary knives. Cutting mills can reduce materials to particles on the order of 100 μm.

Cyclimetric Water   *See* Polywater.

Danish Agar   *See* Furcellaran.

Darcy's Law   *See* Permeability.

Dark-Field Illumination   A kind of illumination for microscopy in which the illumination of a specimen is arranged so that transmitted light falls out of the optical path of the microscope and only light scattered by a dispersed phase is observed. It is used to detect the presence of dispersed species that are smaller than the resolving power of the microscope. Sometimes termed dark-ground illumination. A microscope using this principle is referred to as an ultramicroscope. Example: commonly used in particle microelectrophoresis. Compare to Bright-Field Illumination.

Dark-Field Microscope   *See* Ultramicroscope, Dark-Field Illumination.

Dark-Ground Illumination   *See* Dark-Field Illumination.

Deaeration   The removal of the gas phase from a dispersion. Example: some nonaqueous foams (made from bitumen or heavy crude oils) are very viscous and are deaerated by processes such as contacting with steam in cascading froth, countercurrent steam-flow vessels.

Deborah Number   In rheology, the dimensionless ratio of relaxation time for a process to the time of observation of that process.

Debye Forces   Attractive forces between molecules due to dipole-induced dipole interaction. *See also* London Forces.

Debye–Hückel Parameter   *See* Debye Length.

Debye–Hückel Theory   A description of the behavior of electrolyte solutions in which ions are treated as point charges and their distribution is described in terms of a competition between electrical forces and thermal motion. *See also* DLVO Theory.

Debye Length   A parameter in the Debye–Hückel theory of electrolyte solutions, κ^{-1}. For aqueous solutions at 25 °C, $\kappa = 3.288\sqrt{I}$ in reciprocal nanometers, where I is the ionic strength of the solution. The Debye length is also used in the DLVO theory, where it is referred to as the electric-double-layer thickness and represents the distance over which the potential falls to $1/e$, about one-third, of the value of the surface potential. See Electric- Double-Layer Thickness.

Debye Parameter   *See* Debye Length.

Deflocculation   The reverse of aggregation (or flocculation or coagulation). Peptization means the same thing.

Defoamer   *See* Foam Breaker, Antifoaming Agent.

Degree of Association   In micelles, this is the number of surfactant molecules in the micelle. *See* Aggregation Number.

Degree of Polymerization   The number of repeating monomers in a polymer molecule.

Demulsification   *See* Demulsifier.

Demulsifier   Chemical: Any agent added to an emulsion that causes or enhances the rate of breaking of the emulsion (separation into its constituent liquid phases). Demulsifiers may act by any of a number of different mechanisms, which usually include enhancing the rate of droplet coalescence.

Device: Any device that is used to break emulsions. Such devices may employ chemical, electrical, or mechanical means, or a combination, to break an emulsion and cause separation into its constituent liquid phases.

Dense Nonaqueous-Phase Liquid (DNAPL)   *See* Nonaqueous-Phase Liquid.

Depletion Flocculation   The flocculation of dispersed species induced by the interaction of adsorbed polymer chains. *See* Sensitization.

Depletion Stabilization   The stabilization of dispersed species induced by the interaction (steric stabilization) of adsorbed polymer chains. Also called steric stabilization. *See* Protection.

Depolarization   In light scattering, the reduction in polarization of light scattered at 90° from anisotropic particles as compared to that scattered from isotropic particles. *See also* Cabannes Factor.

Depolarization Ratio   In light scattering, the ratio of intensities of light polarized horizontally to vertically.

Depressant   Any agent that may be used in froth flotation to selectively reduce the effectiveness of collectors for certain mineral components. *See also* Froth Flotation, Activator.

Desalter   An oil-field or refinery apparatus used to separate water and associated dissolved salts from crude oil.

Desorption   The process by which the amount of adsorbed material becomes reduced. That is, the converse of adsorption. Desorption is a different process from negative adsorption. *See also* Adsorption.

Detergency   The action of surfactants that causes or aids in the removal of foreign material from solid surfaces by adsorbing at interfaces and reducing the energy needed to effect the removal. The processes of removal by dissolution and removal by abrasion are not considered to be part of detergency. *See also* Detergent.

Detergent   A surfactant that has cleaning properties in dilute solutions. As commercial cleaning products, detergents are actually formulations containing a number of chemical components, including surfactants, builders, bleaches, brighteners, enzymes, opacifiers, and fragrances. In such formulations there is usually a principal surfactant, termed the main active surfactant, and secondary surfactant(s), termed the coactive surfactant(s).

Detergentless Microemulsion   Does not really refer to an emulsion but rather to making an otherwise insoluble component (an oil) soluble by adding a third component. *See* reference 17.

Dewetting   In antifoaming, the process by which a droplet or particle of antifoaming agent enters the gas–liquid interface and displaces some of the original liquid from the interface. The liquid is usually an aqueous phase, so the process is sometimes referred to as dewetting.

Dialysate   *See* Dialysis.

Dialysis   A separation process in which a colloidal dispersion is separated from a noncolloidal solution by a semipermeable membrane, that is, a membrane that is permeable to all species except the colloidal-sized ones. Osmotic pressure difference across the membrane drives the separation. The solution containing the colloidal species is referred to as the retentate or dialysis residue. The solution that is free of colloidal species is referred to as the dialysate or permeate; at equilibrium (no osmotic pressure difference) this solution is referred to as the equilibrium dialysate. *See also* Ultrafiltration.

Dialysis Residue   *See* Dialysis.

Diamagnetic   A material that is repelled by an external magnetic field. In contrast, paramagnetic materials are attracted into an external magnetic field, and ferromagnetic materials have magnetic properties independently of external magnetic fields.

Dielectric Constant   *See* Permittivity.

Dielectric Saturation   The reduction in relative permittivity due to the application of an electric field.

Dielectrophoresis   The motion of dipolar colloidal species caused by an imposed, nonhomogeneous electric field. The species move toward the most intense part of the electric field with a dielectrophoretic velocity that depends on their dipole moment, any electric charge and the electric field intensity gradient. *See also* Clausius–Mossotti Factor, Levitator.

Dielectrophoretic Force   *See* Dielectrophoresis.

Differential Capacitance of the Electric Double Layer   *See* Capacitance of the Electric Double Layer.

Differential Diffusion Coefficient   *See* Diffusion Coefficient.

Differential Maximum Bubble Pressure Method   *See* Maximum Bubble Pressure Method.

Differential Refractometer   An instrument capable of measuring refractive index differences between materials, such as between a solution and its pure solvent. Such instruments can be capable of determining very small differences in refractive index.

Differential Viscosity   The rate of change of shear stress with respect to shear rate, taken at a specific shear rate ($\eta_D = d\,\tau/d\,\dot\gamma$). *See* Table 4.

Diffuse Double Layer   *See* Diffuse Layer, Electric Double Layer.

Diffuse Layer   The Gouy layer, in an Electric Double Layer.

Diffusion   The spontaneous movement of species in response to a gradient in their chemical potential. Fick's first and second laws specify diffusion coefficients in terms of concentration gradients and form the basis for experimental measurements of diffusion coefficients.

Diffusion Coefficient   According to Fick's first law, the diffusion coefficient (properly the differential diffusion coefficient) is the ratio of amount of species flowing through unit area in unit time, to the concentration gradient of the same species. If extrapolated to zero concentration of diffusing species, it is the limiting diffusion coefficient. The self-diffusion coefficient is the diffusion coefficient of a species when there is no chemical potential gradient.

Diffusiophoresis   The movement of a colloidal species in response to the concentration gradient of another dissolved, noncolloidal, solute.

Diffusivity   *See* Diffusion Coefficient.

Dilatant   A non-Newtonian fluid for which viscosity increases as the shear rate increases. The process is termed shear-thickening.

Dilational Elasticity   *See* Film Elasticity.

Diluent   A low-boiling petroleum fraction, such as naphtha, that is added to a more viscous high- boiling petroleum liquid or oil-continuous emulsion. The diluent is usually added to reduce viscosity.

Dipole   A separation of electrical charge within an electrically neutral species. This is different from bipolar. Dipoles may be permanent or induced temporarily by an applied electric field.

Discontinuous Phase   *See* Dispersed Phase.

Discreteness of Charge   Charged colloidal species usually obtain their charge from a collection of discrete charge groups present at their surfaces. This discreteness of charge is, however, frequently approximated as a uniform surface charge distribution in descriptions of colloidal phenomena (e.g., DLVO theory). *See also* Esin–Markov Effect.

Disjoining Pressure   The negative derivative with respect to distance of the Gibbs energy of interaction per unit area yields a force per unit area between colloidal species, termed the disjoining pressure. Example: in a thin liquid film, the disjoining pressure equals the pressure, beyond the external pressure, that has to be applied to the liquid in the film in order to maintain a given film thickness.

Disk Attrition Mill   *See* Disk Mill.

Disk Mill   A machine for the comminution, or size reduction, of wood products or other material. Such machines crush the input material between two grinding plates mounted on rotating disks. Also termed disk attrition mill.

Dispersant   Any species that may be used to aid in the formation of a colloidal dispersion. Often a surfactant.

Disperse   *See* Dispersion.

Dispersed Phase   In a colloidal dispersion, the phase that is distributed, in the form of particles, droplets, or bubbles, in a second, immiscible phase that is continuous. Also referred to as the disperse, discontinuous, or internal phase. *See also* Continuous Phase.

Disperse Phase   *See* Dispersed Phase.

Dispersing Agent   *See* Dispersant.

Dispersion   Colloids: A system in which finely divided droplets, particles, or bubbles are distributed in another phase. As it is usually used, dispersion implies a distribution without dissolution. An emulsion is an example of a colloidal dispersion; *see also* Colloidal.

Fluid-flow phenomena: The mixing of one fluid in another, immiscible fluid by convection and molecular diffusion during flow through capillary spaces or porous media.

Groundwater contamination: The mixing of a contaminant with a noncontaminant phase. The mixing is due to the distribution of flow paths, tortuosity of flow paths, and molecular diffusion.

Dispersion Forces   Interaction forces between any two bodies of finite mass. Sometimes called van der Waals forces, they include the Keesom orientation forces between dipoles, Debye induction forces between dipoles and induced dipoles, and London (van der Waals) forces between two induced dipoles. Also referred to as Lifshitz–van der Waals forces.

Dispersion Methods   The class of mechanical methods used for preparing colloidal dispersions in which particles or droplets are progressively subdivided. *See also* Condensation Methods.

Dispersion Mill   *See* Colloid Mill.

Dissolved-Gas Flotation   *See* Froth Flotation.

Disymmetry Ratio   In light scattering, the ratio of light scattered at an angle of 45° to that at 135°. The ratio is related to the size and shape of the dispersed species causing the scattering.

Division   *See* Lamella Division.

DLVO Theory   An acronym for a theory of the stability of colloidal dispersions developed independently by B. Derjaguin and L. D. Landau in one research group, and by E. J. W. Verwey and J. Th. G. Overbeek in another. The theory was developed to predict the stability against aggregation of electrostatically charged particles in a dispersion.

DNAPL   Dense nonaqueous-phase liquid. *See* Nonaqueous-Phase Liquid.

Donnan E.M.F.   *See* Donnan Equilibrium.

Donnan Equilibrium   The equilibrium in a system in which one ionic solution is separated from another by a semipermeable membrane or other barrier, not permeable to at least one of the ionic species. The potential difference at zero current between two identical salt bridges placed into the two solutions is the Donnan potential, or Donnan E.M.F. *See also* Membrane Potential.

Donnan Potential   *See* Donnan Equilibrium.

Donnan Pressure   *See* Colloid Osmotic Pressure.

Doppler Broadening   *See* Doppler Effect.

Doppler Effect   The change in frequency of radiation emanating from a source that is in motion relative to the stationary position of detection. Also referred to as Doppler Broadening or Doppler Shift.

Doppler Shift   *See* Doppler Effect.

Dorn Effect   *See* Sedimentation Potential.

Dorn Potential   *See* Sedimentation Potential.

Double Layer   *See* Electric Double Layer.

Double-Layer Thickness   *See* Electric-Double-Layer Thickness.

Drag   The force due to friction experienced by a moving dispersed species.

Draves Wetting Test   A method for comparing the wetting power of surfactants. It measures the time required for complete wetting of a piece of cloth or skein of yarn placed at the surface of a surfactant solution, under specified test conditions. Different systems are compared in terms of their wetting times. *See also* Wetting.

Drizzle   *See* Atmospheric Aerosols of Liquid Droplets, Table 1.

Dropping Mercury Electrode   An electrode comprising a capillary filled with mercury and arranged such that by forming and releasing mercury drops from the capillary tip, a renewable electrode surface can be presented to a solution under study. Used in polarography.

Drop-Volume Method   A method for determining surface or interfacial tension based on measuring the volume of drops that form at and fall from the tip of a capillary. *See also* Drop-Weight Method.

Drop-Weight Method   A method for determining surface or interfacial tension based on measuring the mass of drops that form at and fall from the tip of a capillary. In early literature referred to as the Stalagmometric Method. *See also* Drop-Volume Method.

Dry Foam   *See* Foam.

du Nouy Ring Method   A method for determining surface or interfacial tension based on measuring the force needed to pull an inert ring through an interface. *See also* Wilhelmy Plate Method.

Duplex Film   Any film that is thick enough for each of its two interfaces to be independent of each other and exhibit their own interfacial tensions. A duplex film is thus thicker than a monomolecular film.

Dust   An aerosol of solid particles (dispersion of solid particles in gas) in which the particle sizes are greater than 1 μm in diameter. *See also* Aerosol, Fume.

Dynamic Interfacial Tension   *See* Equilibrium Surface Tension.

Dynamic Surface Tension   *See* Equilibrium Surface Tension.

EACN   *See* Equivalent Alkane Carbon Number.

EDL   *See* Electric Double Layer.

EELS   Electron energy loss spectroscopy. *See* Characteristic Energy-Loss Spectroscopy.

Effective Viscosity   For foams or emulsions flowing in porous media, the foam's effective viscosity is that calculated from Darcy's Law. This is an approximation because foams are compressible, and both foams and emulsions are usually non-Newtonian.

EID   *See* Electron-Impact Desorption Spectroscopy.

Eigenkolloide   *See* Pseudocolloid.

Eilers Equation   An empirical equation for estimating the viscosity of a dispersion. *See* Table 5.

Einstein Equation   Diffusion: Relation between the diffusion coefficient of a dispersed species and its friction factor: $D = kT/f$, where $f = 6\pi\eta r$ (η is viscosity, r is radius).

Viscosity: An equation for estimating the viscosity of a dispersion. *See* Table 5.

Einstein–Smoluchowski Equation   Relation giving the displacement (root mean square) of a dispersed species having

Brownian motion: $\sqrt{<x>} = \sqrt{(2Dt)}$, where D is diffusion coefficient and t is time.

EIS   Electron-impact spectroscopy. *See* Characteristic Energy-Loss Spectroscopy.

Elasticity   The ability of a material to change its physical dimensions when a force is applied to it, and then restore its original size and shape when the force is removed. *See also* Surface Elasticity.

Elasticity Number   A dimensionless quantity (E_s) characterizing the surface-tension gradient in a thinning foam film. For systems containing only one surfactant: $E_s = -(d\gamma/d \ln \rho^s)(R_f/[\eta D])$, where γ is the surface tension, ρ^s is the surface density of the surfactant, R_f is the thin-film radius, η is the bulk liquid viscosity, and D is the diffusivity of the surfactant.

Elastic Limit   The largest force per unit area that can be applied to an elastic substance without causing irreversible deformation.

Elastic Low-Energy Electron Diffraction (ELEED)   *See* Low-Energy Electron Diffraction.

Elastic Scattering   *See* Light Scattering.

Electric Double Layer   (EDL) An idealized description of the distribution of free charges in the neighborhood of an interface. Typically the surface of a charged species is viewed as having a fixed charge of one sign (one layer), while oppositely charged ions are distributed diffusely in the adjacent liquid (the second layer). The second layer may be considered to be made up of a relatively more strongly bound Stern layer in close proximity to the surface and a relatively more diffuse layer (Gouy layer, or Gouy–Chapman layer), at greater distance.

Electric-Double-Layer Thickness   A measure of the decrease of potential with distance in the diffuse part of an electric double layer. It is the distance over which the potential falls to $1/e$, or about one-third, of the value of the surface or Stern layer potential, depending on the model used. Also termed the Debye length.

Electrocapillarity   Refers to the relationship between surface electric potential and surface or interfacial tension. This is most evident for curved interfaces between mercury and aqueous solutions.

The mercury–water interface is normally positively charged. If an electric field is applied to reduce the interfacial electric potential, then the interfacial tension typically rises to a maximum and thereafter decreases. This relationship is known as the electrocapillary curve: The part of the curve where interfacial tension is increasing is referred to as the ascending branch or anodic branch; the part of the curve where interfacial tension is decreasing is referred to as the descending branch or cathodic branch. The maximum in the curve is termed the electrocapillary maximum and occurs at zero net surface electric charge (the zero point of charge).

Electrocapillary Maximum   See Electrocapillarity.

Electrocratic   A dispersion stabilized principally by electrostatic repulsion. *See also* Lyocratic.

Electrocrystallization   A kind of electrodeposition in which ions from solution become deposited on or into an electrode surface and then participate in crystallization, the building up of old crystals, or the growing of new ones at the electrode surface.

Electrodecantation   A separation process in which a colloidal dispersion is separated from a noncolloidal solution by an applied electric field together with the force of gravity. Also called electrophoresis convection.

Electrodeposition   The deposition of dissolved or dispersed species on an electrode under the influence of an electric field.

Electrodialysate   *See* Electrodialysis.

Electrodialysis   A separation process somewhat like dialysis and ultrafiltration, in which a colloidal dispersion is separated from a noncolloidal solution by a semipermeable membrane, that is, a membrane that is permeable to all species except the colloidal-sized ones. Here an applied electric field (rather than osmotic pressure or an applied pressure) across the membrane drives the separation. As in dialysis and ultrafiltration the solution containing the colloidal species is referred to as the retentate or dialysis residue. However, the solution that is free of colloidal species is referred to as electrodialysate rather than dialysate because the composition is usually different from that produced by dialysis. *See also* Dialysis, Ultrafiltration.

Electroendosmosis   *See* Electro-osmosis.

Electroformed Sieve   *See* Particle Size Classification.

Electrokinetic   A general adjective referring to the relative motions of charged species in an electric field. The motions may be either of charged, dispersed species or of the continuous phase, and the electric field may be either an externally applied field or else created by the motions of the dispersed or continuous phases. Electrokinetic measurements are usually aimed at determining Zeta Potentials.

Electrokinetic Potential   *See* Zeta Potential.

Electron Energy Loss Spectroscopy (EELS)   *See* Characteristic Energy-Loss Spectroscopy.

Electron-Impact Desorption Spectroscopy (EID)   A technique for characterizing surfaces and adsorbed species in which a high- energy electron beam is used to cause the ejection of surface ions, whose energy is measured.

Electron-Impact Spectroscopy (EIS)   *See* Characteristic Energy-Loss Spectroscopy.

Electron-Loss Spectroscopy (ELS)   *See* Characteristic Energy-Loss Spectroscopy.

Electron Microscopy   There are three principal types of electron microscopy: transmission electron microscopy (TEM), scanning electron microscopy (SEM), and scanning transmission electron microscopy (STEM). TEM is analogous to transmitted-light microscopy but uses an electron beam rather than light, and uses magnetic lenses to produce a magnified image on a fluorescent screen. In SEM a surface is scanned by a focused electron beam and the intensity of secondary electrons is measured and used to form an image on a cathode-ray tube. In STEM a surface is scanned by a very narrow electron beam that is transmitted through the sample. The intensities in the formed image are related to the atomic numbers of atoms scanned in the sample. *See also* Field Emission Microscopy, Field Ion Microscopy, Scanning Tunneling Microscopy.

Electron Paramagnetic Resonance   *See* Electron Spin Resonance.

Electron Spectroscopy for Chemical Analysis (ESCA)   *See* Photoelectron Spectroscopy.

Electron Spin Resonance Spectroscopy (ESR)   Measurement of the frequency of applied energy needed to induce resonance in the energy levels occupied by electrons. The resonance frequency depends on the local environment of the electrons, hence on molecular structure. Also referred to as electron paramagnetic resonance or EPR spectroscopy.

Electron-Stimulated Desorption Spectroscopy (ESD)   *See* Photon-Stimulated Desorption Spectroscopy.

Electro-osmosis   The motion of liquid through a porous medium caused by an imposed electric field. The term replaces the older terms electrosmosis and electroendosmosis. The liquid moves with an electro-osmotic velocity that depends on the electric surface potential in the stationary solid and on the electric field gradient. The electro-osmotic volume flow is the volume flow rate through the porous plug and is usually expressed per unit electric field strength. The electro-osmotic pressure is the pressure difference across the porous plug that is required to just stop electro-osmotic flow.

Electro-osmotic Pressure   *See* Electro-osmosis.

Electro-osmotic Velocity   *See* Electro-osmosis.

Electro-osmotic Volume Flow   *See* Electro-osmosis.

Electrophoresis   The motion of colloidal species caused by an imposed electric field. The term replaces the older term cataphoresis. The species move with an electrophoretic velocity that depends on their electric charge and the electric field gradient. The electrophoretic mobility is the electrophoretic velocity per unit electric field gradient and is used to characterize specific systems. An older synonym, no longer in use, is kataphoresis. The term microelectrophoresis is sometimes used to indicate electrophoretic motion of a collection of particles on a small scale. Previously, microelectrophoresis was used to describe the measurement techniques in which electrophoretic mobilities are determined by observation through a microscope. The recommended term for these latter techniques is now microscopic electrophoresis (*see* reference 4).

Electrophoresis Convection   *See* Electrodecantation.

Electrophoretic Mobility   *See* Electrophoresis.

Electrophoretic Relaxation   When a charged particle or droplet undergoes electrophoresis, it will tend to move somewhat ahead of its counterions and associated liquid. This movement causes a distortion in the symmetry of the electric double layer and also introduces a local electric field gradient acting in opposition to the external field. This local electric field gradient produces a retarding force on the particle or droplet: the electrophoretic relaxation. *See also* Electrophoretic Retardation.

Electrophoretic Retardation   The effect of an electric field gradient on counterions in the electric double layer surrounding a charged species. When a charged particle or droplet undergoes electrophoresis, its counterions and associated liquid are simultaneously driven to move in the opposite directions. Thus, a retarding force is produced on the particle or droplet: the electrophoretic retardation. *See also* Electrophoretic Relaxation.

Electrophoretic Velocity   *See* Electrophoresis.

Electrosmosis   *See* Electro-osmosis.

Electrostatic Treater   A vessel used to break emulsions by promoting coalescence through the application of an electric field. *See* Treater.

Electrosteric Stabilization   The stabilization of a dispersed species by a combination of electrostatic and steric repulsions. An example is the stabilization of suspended solids by adsorbed polyelectrolyte molecules.

Electroviscous Effect   Any influence of electric double layer(s) on the flow properties of a fluid. The primary electroviscous effect refers to an increase in apparent viscosity when a dispersion of charged colloidal species is sheared. The secondary electroviscous effect refers to the increase in viscosity of a dispersion of charged colloidal species that is caused by their mutual electrostatic repulsion (overlapping of electric double layers). An example of the tertiary electroviscous effect would be for polyelectrolytes in solution where changes in polyelectrolyte molecule conformations and their associated effect on solution apparent viscosity occur.

ELEED   Elastic low-energy electron diffraction. *See* Low-Energy Electron Diffraction.

Ellipsometry   The subject concerned with the behavior of light when it passes through or is reflected by an interface. In ellipsometry an incident beam of plane-polarized light is caused to be reflected from a coated surface and the degree of elliptical polarization produced is measured. This technique allows the thickness of the surface coating to be determined.

Elliptical Jet Method   *See* Oscillating Jet Method.

ELS   Electron-loss spectroscopy. *See* Characteristic Energy-Loss Spectroscopy.

Elutriation   The separation of smaller sized, lighter particles from larger sized, heavier particles due to the flow of surrounding fluid that tends to "carry" the lighter particles.

Embryo   In colloid science, an aggregate of a small number of species. A critical embryo has a size corresponding to maximum Gibbs energy (constant temperature and pressure). A larger embryo is referred to as a homogeneous nucleus. *See* reference 4.

Emollient   An agent that lends a soft texture to skin, hair, or membrane tissues. Used in formulated personal care products such as skin creams and hair conditioners.

Emulsifier   Any agent that acts to stabilize an emulsion. The emulsifier may make it easier to form an emulsion and to provide stability against aggregation and possibly against coalescence. Emulsifiers are frequently but not necessarily surfactants.

Emulsify   *See* Emulsifier.

Emulsion   A dispersion of droplets of one liquid in another, immiscible liquid, in which the droplets are of colloidal or near-colloidal sizes. The term emulsion may also be used to refer to colloidal dispersions of liquid crystals in a liquid. Emulsions were previously referred to as emulsoids, meaning emulsion colloids. *See also* Macroemulsion, Miniemulsion, Microemulsion.

Emulsion Test   In general, emulsion tests range from simple identifications of emulsion presence and volume through to detailed component analyses. The term emulsion test frequently refers simply to the determination of sediments in an emulsion or oil sample. *See* Basic Sediment and Water.

Emulsion Treater   *See* Treater.

Emulsoid   An older term meaning emulsion colloid. *See* Emulsion.

Encapsulation   *See* Microencapsulation.

Energy of Adhesion   *See* Work of Adhesion.

Energy of Cohesion   *See* Work of Cohesion.

Energy of Immersional Wetting   *See* Work of Immersional Wetting.

Energy of Separation   *See* Work of Separation.

Energy of Spreading   *See* Work of Spreading.

Engulfment   The process in which a particle dispersed in one phase is overtaken by an advancing interface and surrounded by a second phase. Example: when a freezing front (the interface between a solid and its freezing liquid phase) overtakes a particle, the particle will either be pushed along by the front or else it will be engulfed by the front, depending on its interfacial tensions with the solid and with the liquid. *See also* Freezing Front Method.

Enhanced Oil Recovery   The third phase of crude oil production, in which chemical, miscible or immiscible gas, or thermal methods are applied to restore production from a depleted reservoir. Also known as tertiary oil recovery. *See* Primary Oil Recovery, Secondary Oil Recovery.

Entering Coefficient   A measure of the tendency for an insoluble agent to penetrate, or "enter", an interface (usually gas–liquid or liquid–liquid). It is –1 times the Gibbs free energy change for this process, so that entering is thermodynamically favored if the entering coefficient is greater than zero. In a gas–liquid system containing such an agent A, a liquid L, and gas, the entering coefficient is given by $E = \gamma^{\circ}_L + \gamma_{L/A} - \gamma^{\circ}_A$ where γ°_L and γ°_A are surface tensions and $\gamma_{L/A}$ is interfacial tension. When equilibria at the interfaces are not achieved instantaneously, reference is frequently made to the initial entering coefficient and final (equilibrium) entering coefficient. *See also* Spreading Coefficient.

Enthalpy Stabilization   *See* Steric Stabilization.

Entropy Stabilization   *See* Steric Stabilization.

Eötvös Equation   A relation for predicting the variation of surface tension with temperature: $\gamma(M/\rho)^{2/3} = k(T_c - T)$, where M is the molecular mass, ρ is the density, T_c is the critical temperature of the liquid, and k is termed the Eötvös constant. *See* Table 8.

Epitaxy   The growth of crystalline material on the surface of a different material where the substrate material orients the new crystal growth.

EPR   *See* Electron Spin Resonance.

Equation of Capillarity   *See* Young–Laplace Equation.

Equilibrium Contact Angle   The contact angle that is measured when all contacting phases are in equilibrium with each other. The term arises because either or both of the advancing or receding contact angles may differ from the equilibrium value. It is essential to state which interfaces are used to define the contact angle. *See also* Contact Angle.

Equilibrium Dialysate   *See* Dialysis.

Equilibrium Film   *See* Fluid Film.

Equilibrium Interfacial Tension   *See* Equilibrium Surface Tension.

Equilibrium Spreading Coefficient   *See* Spreading Coefficient.

Equilibrium Surface Tension   Surface or interfacial tensions may change dynamically as a function of the age of the surface or interface. Thus the dynamic (pre-equilibrium) tensions are distinguished from the limiting, or equilibrium, tensions.

Equivalent Alkane Carbon Number (EACN)   Each surfactant, or surfactant mixture, in a reference series will produce a minimum interfacial tension (IFT) when measured against a different *n*-alkane. For any crude oil or oil component, a minimum IFT will be observed against one of the reference surfactants. The EACN for the crude oil refers to the *n*-alkane that would yield minimum IFT against that reference surfactant. The EACN thus allows predictions to be made about the interfacial tension behavior of a crude oil in the presence of surfactant. *See* references 29 and 30.

Equivalent Film Thickness   Refers to an experimentally determined fluid film thickness; the term equivalent refers to certain assumptions about the structure and properties of the film that have been made. The experimental technique used should also be stated when using this term.

Equivalent Spherical Diameter   The diameter of a sedimenting species determined from Stokes' law assuming a spherical shape. Also referred to as the Stokes diameter or (divided by a factor of 2) the settling radius.

ESCA   Electron spectroscopy for chemical analysis. *See* Photoelectron Spectroscopy.

ESD   Electron-stimulated desorption spectroscopy. *See* Photon-Stimulated Desorption Spectroscopy.

Esin–Markov Effect   The change in zero point of charge of a species that occurs when the electrolyte can become specifically adsorbed. In the presence of indifferent electrolytes, the zero point of charge is a constant.

ESR Spectroscopy   *See* Electron Spin Resonance.

Evanescent Foam   A transient foam that has no thin-film persistence and is therefore very unstable. Such foams exist only where new bubbles can be created faster than existing bubbles rupture. Examples: air bubbles blown rapidly into pure water; the foam created when a champagne bottle is opened.

EXAFS   *See* Extended X-Ray Absorption Fine Structure Spectroscopy.

Excess Quantities   *See* Gibbs Surface.

Excluded Volume   The volume in a system, or near an interface, that is not accessible to molecules or dispersed species because of the presence of other species in that volume. *See also* Free Volume.

Expansion Factor   In foaming, the ratio of foam volume produced to the volume of liquid used to make the foam. Also termed the expansion ratio.

Expansion Ratio   *See* Expansion Factor.

Extended X-Ray Absorption Fine Structure Spectroscopy (EXAFS)   A technique for studying the separation distance of surface atoms; it is related to X-ray photoelectron spectroscopy and is based on the effect of backscattering. X-ray absorption is determined as a function of the energy of the incident X-ray beam. A related technique is surface-extended X-ray absorption fine structure spectroscopy (SEXAFS). *See also* Table 9.

External Phase   *See* Continuous Phase.

External Surface   When a porous medium can be described as consisting of discrete particles, the outer surface of the particles is termed the external surface. *See also* Internal Surface.

Extinction Coefficient   *See* Absorbance.

Extra-Heavy Crude Oil   A naturally occurring hydrocarbon having a viscosity less than 10,000 mPa·s at ambient deposit temperature, and a density greater than 1000 kg/m^3 at 15.6 °C. *See* references 25–27.

Fatty Acid Soaps A class of surfactants comprising the salts of aliphatic carboxylic acids having hydrocarbon chains of between 6 and 20 carbon atoms. Fatty acid soaps are no longer restricted to molecules having their origins in natural fats and oils.

Fatty Alcohol Surfactants The class of primary alcohol surfactants having hydrocarbon chains of between 6 and 20 carbon atoms. Fatty alcohol surfactants are no longer restricted to molecules having their origins in natural fats and oils.

FDS Flash desorption spectroscopy. *See* Temperature-Programmed Reaction Spectroscopy.

FEM *See* Field Emission Microscopy.

Feret's Diameter A statistical particle diameter; the length of a line drawn parallel to a chosen direction and taken between parallel planes drawn at the extremities on either side of the particle. This diameter is thus the maximum projection of the particle onto any plane parallel to the chosen direction. The value obtained depends on the particle orientation, thus these measurments have significance only when a large enough number of measurements are averaged together. *See also* Martin's Diameter.

Ferromagnetic A material that has magnetic properties (below the Curie point) independently of external magnetic fields. In contrast, paramagnetic materials acquire magnetic properties when placed in an external magnetic field and become attracted into the field, whereas diamagnetic materials are weakly repelled by an external magnetic field.

Fick's First Law   *See* Diffusion Coefficient.

Field Emission Microscopy (FEM)   A type of electron microscopy in which electrons are emitted from the charged hemispherical tip of a metal wire. The electron beam is detected at a hemispherical fluorescent screen and used to form a highly magnified image that can be used to elucidate the crystal structure of the metal tip. *See also* Electron Microscopy, Field Ion Microscopy, Table 9.

Field Ion Microscopy   A variation of field emission microscopy in which gas molecules in the vicinity of a positively charged, fine metal tip lose an electron. The resulting positive ions accellerate away to strike a fluorescent screen where they are detected and used to form an image of the crystal structure of the metal tip. In this technique individual atoms can be resolved. *See also* Electron Microscopy, Field Emission Microscopy, Table 9.

Filler   Fine-grained, inert material that is added to paper, paint, rubber, resin, etc., to improve their properties in some way.

Film   Any layer of material that covers a surface and is thin enough to not be significantly influenced by gravitational forces. *See also* Monolayer Adsorption, Duplex Film.

Film Balance   A shallow trough that is filled with a liquid, and on top of which is placed material that may form a monolayer. The surface area available can be adjusted by moveable barriers, and, by means of a float, any surface pressure thus created can be measured. Also called Langmuir film balance, Langmuir trough, and Pockels–Langmuir–Adam–Wilson–McBain trough or PLAWM trough.

Film Compressibility   The ratio of relative area change to differential change in surface tension. *See also* Film Elasticity.

Film Drainage   The drainage of liquid from a lamella of liquid separating droplets or bubbles of another phase (i.e., in a foam or emulsion). Also termed thin-film drainage. *See also* Fluid Film.

Film Elasticity   The differential change in surface tension with relative change in area. Also termed surface elasticity, dilational elasticity, areal elasticity, compressional modulus, surface dilational modulus, or modulus of surface elasticity. For fluid films the surface tension of one surface is used. The Gibbs film (surface) elasticity is the equilibrium value. If the surface tension is dynamic (time-dependent) in character then, for nonequilibrium values, the term Marangoni film (surface) elasticity is used. The compressibility of a film is the inverse of the film elasticity.

Film Element   Any small, homogeneous region of a thin film. The film element includes the interfaces.

Film Flotation Technique   *See* Hydrophobic Index.

Film Pressure   The pressure, in two dimensions, exerted by an adsorbed monolayer. It is formally equal to the difference between the surface tension of pure solvent and that of the solution of adsorbing solute. It can be measured by using the film balance. *See also* Film Balance.

Film Tension   An expression of surface tension applied to thin liquid films that have two equivalent surfaces. The film tension is twice the surface tension.

Film Water   In soil science, the film of water that remains, surrounding soil particles, after drainage. This layer may range from several to hundreds of molecules thick and comprises water of hydration plus water trapped by capillary forces.

Filter Ripening   In water filtration, the process in which deposition of an initial layer of particles causes the filter surface to take on a nature more similar to the particles to be removed. This process enhances the filtering (hence, removal) of the particles.

Filtration   The process of removing particles or large molecules from a fluid phase by passing the fluid through some medium that will not permit passage of the particles or large molecules. The filtration medium may comprise woven fabric or metal fibers, porous media, or other materials. In water treatment, filtration refers to sand or mixed-bed granular filters that are used to remove colloidal and larger sized particles.

FIM   *See* Field Ion Microscopy.

Final Spreading Coefficient   *See* Spreading Coefficient.

Fine Sand   *See* Sand, Table 3.

Flash Desorption Spectroscopy (FDS)   *See* Temperature-Programmed Reaction Spectroscopy.

Floc   *See* Flocculation.

Flocculation   *See* Aggregation. The products of the flocculation process are referred to as flocs.

Flocculation Value   *See* Critical Coagulation Concentration.

Flory–Huggins Theory   A description of polymer solutions in which the polymer molecules adopt random coil configurations. The polymer groups or solvent molecules are taken to occupy positions in a lattice. This arrangement allows calculation of the thermodynamics of polymer mixing in solutions.

Flory–Krigbaum Theory   A description of nonideal polymer solutions; it statistically describes what happens as polymer coils approach each other. This description of steric stabilization has also been applied to other dispersions, such as suspensions.

Flory Point   *See* Theta Temperature.

Flory Temperature   *See* Theta Temperature.

Flotation   *See* Froth Flotation, Sedimentation.

Flow-Line Treating   In oil production or processing, the process in which emulsion is continuously broken and separated into oil and water bulk phases. This process is as opposed to batch treating of emulsions. *See also* Treater.

Fluid Film   A thin-fluid phase, usually of thickness less than about 1 μm. Such films may be specified by abbreviations similar to those used for emulsions; for example, some common designations are

> A/W/A for a water film in air
> W/O/W for an oil film in water
> O/W/O for a water film in oil
> W/O/A for an oil film between water and air.

Fluid films are usually unstable to breakage that is due to rupture: thinning to the point of allowing contact of the separating phase(s). There may, however, be film thicknesses at which a film is stable or metastable to thickness changes. Films with this property are equilibrium films. Otherwise fluid films may be distinguished by rapid (mobile film) or slow (rigid film) thickness changes. *See also* Black Film.

Fluorescence Microscopy   Light microscopy in which ultraviolet light is used to induce fluorescence in a specimen. The fluorescent light is then used to form the magnified image, in either transmitted- or reflected-light modes (the ultraviolet light is filtered out at this stage).

Fluorescent Whitening Agents (FWA)   *See* Optical Brighteners.

Fluorochroming   The use of fluorescent dye(s) to stain a specimen and make it visible for microscopic study.

Flux   The flow rate of matter or energy per unit area.

Fly Knives   The rotating cutting blades in a cutting mill machine for comminution.

Foam   A dispersion of gas bubbles in a liquid, in which at least one dimension falls within the colloidal size range. Thus a foam typically contains either very small bubble sizes or, more commonly, quite large gas bubbles separated by thin liquid films. The thin liquid films are called lamellae (or laminae). Sometimes distinctions are drawn as follows. Concentrated foams, in which liquid films are thinner than the bubble sizes and the gas bubbles are polyhedral, are termed polyederschaum. Low-concentration foams, in which the liquid films have thicknesses on the same scale or larger than the bubble sizes and the bubbles are approximately spherical, are termed gas emulsions, gas dispersions, or kugelschaum. *See also* Evanescent Foam, Froth, Aerated Emulsion, Solid Foam.

Foam Booster   *See* Foaming Agent.

Foam Breaker   Any agent that acts to reduce or eliminate foam stability. Also termed defoamer. A more general term is antifoaming agent. *See also* Antifoaming Agent.

Foam Drainage   The drainage of liquid from liquid lamellae separating bubbles in a foam. *See also* Fluid Film.

Foam Emulsion   *See* Aerated Emulsion.

Foamer   *See* Foaming Agent.

Foam Flooding   Enhanced oil recovery: The process in which a foam is made to flow through an underground reservoir. The foam, which may be either generated on the surface and injected or generated in situ, is used to increase the drive fluid viscosity and improve its sweep efficiency.

Petroleum processing: In refinery distillation and fractionation towers the occurrence of foams, which can carry liquid into regions of the towers intended for vapor.

Foam Fractionation   A separation method in which a component of a liquid that is preferentially adsorbed at the liquid–gas interface is removed by foaming the liquid and collecting the foam produced. Foaming surfactants can be separated in this manner.

Foaming Agent   Any agent that acts to stabilize a foam. The foaming agent may make it easier to form a foam or provide stability against coalescence. Foaming agents are usually surfactants. Also termed foam booster, whipping agent, and aerating agent.

Foaming Power   *See* Increase of Volume upon Foaming.

Foam Inhibitor   Any agent that acts to prevent foaming. Also termed foam preventative. A more general term is antifoaming agent. *See also* Antifoaming Agent.

Foam Number   A relative drainage rate test for foams in which a foam is formed in a vessel and thereafter the remaining foam volume determined as a function of time. The foam number is the volume of bulk liquid that has separated after a specified time interval, expressed as a percentage of the original volume of liquid foamed.

Foamover   In an industrial process vessel, unwanted foam may occasionally build up to such an extent that it becomes carried out the top of the vessel ("foamover") and on to the next part of the process. This carry over of foam and any entrained material that comes with it is frequently detrimental to other parts of a processing operation.

Foam Preventative   *See* Foam Inhibitor, Antifoaming Agent.

Foam Quality   The gas volume fraction in a foam. Expressed as a percentage this is sometimes referred to as Mitchell foam quality. In three-phase systems other measures are used. For example, when foams are formulated to contain solid particles as well, the slurry quality, Q_s, which gives the volume fraction of gas plus solid, may be used: $Q_s = (V_g + V_s)/(V_g + V_s + V_l)$, where V_g, V_s, and V_l denote the volumes of gas, solid, and liquid phases, respectively.

Foam-Stimulation Fluid   A foam, aqueous or nonaqueous, that is injected into a petroleum reservoir to improve the productivity of oil- or gas-producing wells. Some mechanisms of action for foam-stimulation fluids include fracturing, acidizing to increase permeability, and diversion of flow.

Foam Texture   The bubble size distribution in a foam. For foams in porous media, it may be expressed in terms of the length scale of foam bubbles as compared to that for the spaces confining the foam. When the length scale of the confining space is comparable to or less than the length scale of the foam bubbles, the foam is sometimes termed lamellar foam, to distinguish it from the opposite case, termed bulk foam.

Fog   *See* Atmospheric Aerosols of Liquid Droplets, Table 1.

Foreign Colloid   *See* Pseudocolloid.

Forward Scattering   *See* Light Scattering.

Fowkes Equation   A means for estimating the interfacial tension between two liquids, based on the surface tensions and molecular properties of each liquid. *See* Table 8.

Fowler–Guggenheim Equation   An extension of the Langmuir isotherm equation that allows for lateral interactions among adsorbed molecules.

Fractal   A structure that has an irregular shape under all scales of measurement. The fractal dimension of a species is the exponent D to which a characteristic dimension must be raised to obtain proportionality with the overall size of the species. Fractal dimensions are used for species having a dimensionality of between 2 and 3, such as many particle aggregates.

Fractal Dimension   *See* Fractal.

Fradkina Equation   For predicting the relative permittivity of emulsions. *See* Table 7.

Free-Draining Polymer   Polymer molecules in their extended rather than coiled configuration, so that laminar flow of solvent can occur along the molecules.

Free Energy   A measure of the balance of energetic and entropic forces in a system. For systems maintained at constant pressure, the free energy is referred to as the Gibbs free energy (now frequently termed Gibbs energy); $G = H - TS$, where H is the enthalpy, T is temperature, and S is entropy. For systems maintained at constant volume, the free energy is referred to as the Helmholtz free energy (Helmholtz energy); $A = E - TS$, where $E = H - PV$ (P is pressure and V is volume).

Free Molecules   In polymer- or surfactant-containing systems, the molecules of polymer or surfactant dissolved in solution, that is, those that are not adsorbed or precipitated. For surfactant solutions, free surfactant includes those molecules present in micelles.

Free Polymer   *See* Free Molecules.

Free Surfactant   *See* Free Molecules.

Free Volume   The volume in a system, or near an interface, that is available and not occupied by other molecules or dispersed species. *See also* Excluded Volume.

Free Water   The readily separated, nonemulsified water that is coproduced with oil from a production well.

Free-Water Knockout (FWKO)   A vessel designed to separate readily separated (nonemulsified or "free") water from oil or an oil-containing emulsion. Further water and solids removal may be accomplished in a treater.

Freeze-Fracture Method   A sample preparation technique used in electron microscopy in which specimens are quickly frozen in a cryogen, then cleaved to expose interior surfaces. In some techniques the sample is then observed directly in an electron microscope equipped with a cryogenic stage; in other cases the cleaved sample is coated with a metal coating to produce a replica, which is observed in the electron microscope. *See also* Scanning Electron Microscopy, Transmission Electron Microscopy, Replica.

Freezing Front Method   *See* Solidification Front Method.

Fremdkolloide   *See* Pseudocolloid.

Frenkel–Halsey–Hill Isotherm   An adsorption isotherm that takes account of multilayer adsorption on porous substrates.

Freundlich Isotherm   An empirical adsorption isotherm equation for heterogeneous surfaces. The amount adsorbed per mass of adsorbent increases with increasing equilibrium solute concentration at all concentrations and does not exhibit a limiting adsorption value. *See* Adsorption Isotherm.

Fricke Equation   An equation for predicting the conductivity of dispersions. *See* Table 6.

Friction   The surface resistance of a body to motion. The coefficient of friction is given by the frictional force needed to move one surface over another divided by the load normal to the direction of motion along the surfaces (Amonton's law). The static coefficient of friction is that involving the force needed to initiate motion, and the kinetic coefficient of friction involves the force needed to maintain a given rate of motion. The science of friction and lubrication is known as tribology.

Friction Factor   In the rheology of a dispersion, the friction factor relates to the dissipation of energy due to friction at the surfaces of the dispersed species (i.e., due to drag).

Froth   A type of foam in which solid particles are also dispersed in the liquid (in addition to the gas bubbles). The solid particles may even be the stabilizing agent. The term froth is sometimes used to refer simply to a concentrated foam, but this usage is not preferred.

Frother   *See* Frothing Agent.

Froth Flotation   A separation process utilizing flotation, in which particulate matter becomes attached to gas (foam) bubbles. The flotation process produces a product layer of concentrated particles in foam termed froth. Variations include dissolved-gas flotation, in which gas is dissolved in water that is added to a colloidal dispersion. As microbubbles come out of solution they attach to and float the colloidal species. *See also* Scavenging Flotation, Activator, Collector, Depressant.

Frothing Agent   Any agent that acts to stabilize a froth. May make it easier to form a froth and provide stability against coalescence. Frothing agents are usually surfactants. Analogous to foaming agent.

Fulvic Acids   *See* Humic Substances.

Fume   An aerosol of solid particles (dispersion of solid particles in gas) in which the particle sizes are less than 1 μm in diameter. *See also* Aerosol.

Furcellaran   A water-soluble mixture of sulfated galactans derived from seaweeds. Also termed Danish agar. Furcellaran sols are similar to those of agar and can be quite viscous; can readily form gels; and may be used to stabilize certain suspensions, foams, and emulsions. Furcellaran is used in many different applications, especially in foods and medicines. *See also* reference 20, Seaweed Colloids.

FWA   Fluorescent whitening agents. *See* Optical Brighteners.

FWKO   *See* Free-Water Knockout.

Galvani Potential   *See* Inner Potential.

Gas Aphrons   *See* Microgas Emulsions.

Gas Dispersion   *See* Foam, Solid Foam, Gas Emulsion.

Gas Emulsion   "Wet" foams in which the liquid lamellae have thicknesses on the same scale or larger than the bubble sizes. Typically in these cases the gas bubbles have spherical rather than polyhedral shape. Other synonyms include gas dispersion and kugelschaum. If the bubbles are very small and have a significant lifetime, the term microfoam is sometimes used. In petroleum production the term is used to specify crude oil that contains a small volume fraction of dispersed gas. *See also* Foam.

Gegenion   *See* Counterion.

Gel   A suspension or polymer solution that behaves as an elastic solid rather than a liquid. A dried-out gel is termed a xerogel. Examples: gels of gelatin solutions or of clay suspensions.

Gel Filtration Chromatography   *See* Chromatography.

Gel Permeation Chromatography   *See* Chromatography.

Generalized Plastic Fluid   A fluid characterized by both of the following: the existence of a finite shear stress that must be applied before flow begins (yield stress), and pseudoplastic flow at higher shear stresses. *See also* Bingham Plastic Fluid.

Germs   (Nucleation) *See* Nuclei.

***g*-Forces**   *See* Relative Centrifugal Force.

Gibbs Adsorption   *See* Gibbs Surface.

Gibbs Adsorption Isotherm   *See* Gibbs Isotherm.

Gibbs Dividing Surface   *See* Gibbs Surface.

Gibbs–Duhem Equation   The thermodynamic relationship between the partial molar properties of one component in a system and the partial molar properties of other components as a function of composition.

Gibbs Effect   The decrease in surface or interfacial tension that occurs as surfactant concentration increases towards the critical micelle concentration.

Gibbs Elasticity   *See* Film Elasticity.

Gibbs Energy   Gibbs free energy. *See* Free Energy.

Gibbs Energy of Attraction   When two dispersed-phase species approach, they may attract each other as a result of such forces as the London–van der Waals forces. The Gibbs energy of attraction may be thought of as the difference between Gibbs attractive energies of the system at a specified separation distance and at infinite separation. Although IUPAC (reference 4) has discouraged the use of the synonyms potential energy of attraction and attractive potential energy, they are still in common usage. *See also* Gibbs Energy of Interaction, Gibbs Energy of Repulsion.

Gibbs Energy of Interaction   When two dispersed-phase species approach, they experience repulsive and attractive forces such as electrostatic repulsion and van der Waals attraction. The Gibbs energy of interaction may be thought of as the difference between Gibbs energies of the system at a specified separation distance and at infinite separation. An example of the dependence of Gibbs energy of interaction and distance of separation is that calculated from DLVO theory. Although IUPAC (reference 4) has discouraged the use of the synonyms potential energy of interaction and total potential energy of interaction, they are still in common usage. *See also* Gibbs Energy of Attraction, Gibbs Energy of Repulsion, DLVO Theory, Primary Minimum.

Gibbs Energy of Repulsion   When two dispersed-phase species approach, they may repel each other as a result of such forces as electrostatic repulsion. The Gibbs energy of repulsion may be thought of as the difference between Gibbs repulsive energies of the system at a specified separation distance and at infinite separation. Although IUPAC (reference 4) has discouraged the use of the synonyms potential energy of repulsion and repulsive potential energy, they are still in common usage. *See also* Gibbs Energy of Attraction, Gibbs Energy of Interaction.

Gibbs Film Elasticity   *See* Film Elasticity.

Gibbs Free Energy   Now frequently termed Gibbs energy. *See* Free Energy.

Gibbs Isotherm   An equation that relates the Gibbs surface excess (amount adsorbed) to the change in interfacial tension with activity of the adsorbing species.

Gibbs–Marangoni Effect   The effect in thin liquid films and foams whereby stretching an interface causes the surface excess surfactant concentration to decrease, hence surface tension to increase (Gibbs effect); the surface tension gradient thus created causes liquid to flow toward the stretched region, thus providing both a "healing" force and also a resisting force against further thinning (Marangoni effect). Sometimes referred to simply as the Marangoni effect.

Gibbs Phase Rule   *See* Phase Rule.

Gibbs–Plateau Border   *See* Plateau Border.

Gibbs Ring   *See* Plateau Border.

Gibbs Surface   A geometrical surface chosen parallel to the interface and used to define surface excess properties such as the extent of adsorption. *See* Gibbs Surface Excess.

Gibbs Surface Concentration   The Gibbs surface excess adsorption amount divided by the area of the interface.

Gibbs Surface Elasticity   *See* Film Elasticity.

Gibbs Surface Excess   The excess amount of a component actually present in a system over that present in a reference system of the same volume as the real system, and in which the bulk concentrations in the two phases remain uniform up to the Gibbs dividing surface. The terms surface excess concentration or surface excess have now replaced the earlier term superficial density.

Girifalco–Good Equation   An equation for estimating interfacial tension based on the surface tensions and molecular volumes of the two liquids. *See* Table 8.

Girifalco–Good–Fowkes–Young Equation   An equation relating contact angle through a liquid, between a gas and a solid, to the components of surface tension of the solid and liquid that are due to dispersion (van der Waals) forces and the surface tension of the liquid.

Glass-Transition Temperature   The temperature at which a polymer changes from a viscous or elastic state into a nonelastic solid state.

Glue   *See* Adhesive.

Gouy–Chapman Layer   *See* Electric Double Layer.

Gouy–Chapman Theory   A description of the electric double layer in a colloidal dispersion in which one layer of charge is assumed to exist as a uniform charge distribution over a surface, and the counterions are treated as point charges distributed throughout the continuous, dielectric phase.

Gouy Layer   *See* Electric Double Layer.

Granule   Sometimes used to describe particles having sizes greater than about 2000 μm, depending on the classification system used. Also called gravel. *See* Table 3.

Gravel   *See* Granule.

Gravity Separator   *See* Separator.

gs   *See* Relative Centrifugal Force.

Gum   Any hydrophilic plant material, or derivative, that forms a viscous dispersion or solution with water. Example: gum arabic (acacia gum) is derived from Acacia trees and is used in paints, inks, adhesives, and textiles.

Gum Arabic   *See* Gum.

Gun Barrel   A type of settling vessel used to separate water and oil from an emulsion. Typically, a heated emulsion is treated with demulsifier and introduced into the gun barrel where water settles out and is drawn off. Any produced gas is also drawn off.

Guth–Gold–Simha Equation   An empirical equation for estimating the viscosity of a dispersion. *See* Table 5.

Half-Colloid   *See* Lyophilic Colloid.

Half-Micelle   *See* Hemimicelle.

Hamaker Constant   In the description of the London–van der Waals attractive energy between two dispersed bodies, such as particles or droplets, the Hamaker constant is a proportionality constant characteristic of the particle composition. It depends on the internal atomic packing and polarizability of the droplets. Also termed the van der Waals–Hamaker constant. *See also* van der Waals Attraction Constant.

Hammer Mill   A device for reducing the particle size of a solid, for example a pigment, that uses centrifugal force to drive the solid between rotating "hammers" and a stationary ring-shaped "anvil".

Hanging Bubble Method   *See* Pendant Drop Method.

Hanging Drop Method   *See* Pendant Drop Method.

Hanai Equation   An equation for predicting the conductivities or relative permittivities of dispersions. *See* Tables 6 and 7.

Harkins–Jura Isotherm   An adsorption isotherm equation that accounts for the possibility of multilayer adsorption. *See also* Adsorption Isotherm.

Hatschek Equation   An equation for predicting the viscosity of emulsions. *See* Table 5.

Haze   An aerosol of solid particles (dispersion of solid particles in gas) in which the particle sizes are smaller than can be seen without the aid of a microscope. *See also* Aerosol, Fume, Dust.

HDC   *See* Hydrodynamic Chromatography.

Head Group   The lyophilic functional group in a surfactant molecule. In aqueous systems the polar group of a surfactant. *See also* Surfactant, Surfactant Tail.

Heater Treater   *See* Treater.

Heavy Crude Oil   A naturally occurring hydrocarbon having a viscosity less than 10,000 mPa·s at ambient deposit temperature, and a density between 934 and 1000 kg/m^3 at 15.6 °C. *See* references 25–27.

HEED   *See* High-Energy Electron Diffraction.

HEIS   High-energy ion scattering. *See* Ion-Scattering Spectroscopy.

Helmholtz Condenser   *See* Helmholtz Double Layer.

Helmholtz Double Layer   A simplistic description of the electric double layer as a condenser (the Helmholtz condenser) in which the condenser plate separation distance is the Debye length. The Helmholtz layer is divided into an inner Helmholtz plane (IHP) of adsorbed, dehydrated ions immediately next to a surface, and an outer Helmholtz plane (OHP) at the center of a next layer of hydrated, adsorbed ions just inside the imaginary boundary where the diffuse double layer begins. That is, both Helmholtz planes are within the Stern layer.

Helmholtz Energy   Helmholtz free energy. *See* Free Energy.

Helmholtz Free Energy   Now frequently termed Helmholtz energy. *See* Free Energy.

Helmholtz Plane   *See* Helmholtz Double Layer.

Helmholtz–Smoluchowski Equation   *See* Smoluchowski Equation.

Hemacytometer   A type of particle or droplet sizing and counting chamber used in microscopy. The chamber contains an accurately ruled grid of squares and, with the cover slip in place, holds a specified volume of typically 0.1 μL. Originally developed for and usually used for counting blood cells, it is also used for particle or droplet counting and sizing in other suspensions and emulsions.

Hemimicelle   An aggregate of adsorbed surfactant molecules that may form beyond monolayer coverage, the enhanced adsorption being due to hydrophobic interactions between surfactant tails. Hemimicelles (half-micelles) have been considered to have the form of surface aggregates, or of a second adsorption layer with reversed orientation, somewhat like a bimolecular film. In bilayer surfactant adsorption, the term admicelles has also been used (reference 31). Admicellar chromatography and admicellar catalysis make use of media bearing admicelles.

Henry Isotherm   A sorption isotherm describing the linear (with respect to concentration) partitioning of a chemical between phases. Also used to describe adsorption at low surface coverage when a linear approximation is appropriate (in the limit of very low surface coverage).

Henry Equation   A relation expressing the proportionality between electrophoretic mobility and zeta potential for different values of the Debye length and size of the species. *See also* Smoluchowski Equation, Hückel Equation, Electrophoresis.

Heterocoagulation   The coagulation of dispersed species of different types or having different states of surface electric charge.

Heterodisperse   A colloidal dispersion in which the dispersed species (droplets, particles, etc.) do not all have the same size. Subcategories are paucidisperse (few sizes) and polydisperse (many sizes). *See also* Monodisperse, Number-Average Quantities, Mass-Average Quantities.

Heterogeneous Nucleation   *See* Condensation Methods.

Heterovalent   A term that refers to species having different valencies, as opposed to homovalent for the same valency. In ion exchange it means that the adsorbing and desorbing species have different charges.

High-Energy Electron Diffraction (HEED)   A diffraction technique in which a high-energy electron beam is used. The electrons have high penetrating power, so a grazing angle of incidence is used so that measuring the diffraction pattern will yield information about surface structure. The inelastically scattered electrons, having lower energy, are stopped by grids, and the elastically scattered electrons, having the original energy level, are used to form the image pattern. Derived techniques include reflection high-energy electron diffraction (RHEED) and scanning high-energy electron diffraction (SHEED). *See also* Low-Energy Electron Diffraction, Table 9.

High-Energy Ion Scattering (HEIS)   *See* Ion-Scattering Spectroscopy.

Higher Order Tyndall Spectra (HOTS)   The spectra obtained for light scattered at various angles from an illuminated dispersion. *See also* Tyndall Scattering.

High-Resolution Electron Energy Loss Spectroscopy (HREELS)   *See* Characteristic Energy-Loss Spectroscopy.

HIOC   *See* Hydrophobic Ionogenic Organic Compound.

HLB Scale   *See* Hydrophile–Lipophile Balance.

HLB Temperature   *See* Phase Inversion Temperature.

HOC   *See* Hydrophobic Organic Contaminant.

Hofmeister Series   *See* Lyotropic Series.

Homogeneous Nucleation   *See* Condensation Methods.

Homogeneous Nucleation Temperature   The temperature below which the rate of a nucleation process increases rapidly. In practice a narrow range of temperatures represents the transition from very slow to very rapid nucleation. *See also* Condensation Methods.

Homogeneous Nucleus   *See* Embryo.

Homogenizer   Any machine for preparing colloidal systems by dispersion. Examples: colloid mill, blender, ultrasonic probe.

Homopolymer   *See* Polymer.

Homovalent   Species having the same valencies, as opposed to heterovalent for different valencies. In ion exchange it means that the adsorbing and desorbing species have the same charge.

Hooke's Law   Expression of the relation between the force per area applied to an elastic body and the tensile stress, where Young's modulus is the constant of proportionality between them.

HREELS   High-resolution electron energy loss spectroscopy. *See* Characteristic Energy-Loss Spectroscopy.

Hückel Equation   A relation expressing the proportionality between electrophoretic mobility and zeta potential for the limiting case of a species that can be considered small and with a thick electric double layer. *See also* Smoluchowski Equation, Henry Equation, Electrophoresis.

Humic Acids   *See* Humic Substances.

Humic Substances   Polyaromatic and polyelectrolytic organic acids of high molecular mass (about 800–4000 g/mol or higher) that occur in natural water bodies, soils, and sediments. Although significantly aromatic, these acids may have an appreciable aliphatic component as well and may be surface-active (reference 18). Humic substances are operationally divided into humic acids and fulvic acids on the basis of solubility as follows: humic acids are water-soluble above pH 2 but water-insoluble below pH 2; fulvic acids are water-soluble at all pH levels.

Hydrocolloid   Any of the hydrophilic colloidal materials used (mostly) in food products as emulsifying, thickening, and gelling agents. These are mostly carbohydrate polymers, although some are proteins. Examples: agar, carrageenan, dextran, gelatin, guar gum.

Hydrodynamic Chromatography (HDC)   A method of determining the size distribution of dispersed species. A dispersion is made to flow through a packed bed of small beads. Fractions of effluent contain different components according to their rate of travel through the bed. The method is different from gel permeation chromatography in that here the packed beads are not porous. *See* Chromatography.

Hydrophile–Lipophile Balance (HLB scale)   An empirical scale categorizing surfactants in terms of their tendencies to be mostly oil-soluble or water-soluble, hence their tendencies to promote W/O or O/W emulsions, respectively. *See also* Phase Inversion Temperature.

Hydrophilic   A qualitative term referring to the water-preferring nature of a species (atom, molecule, droplet, particle, etc.). For emulsions hydrophilic usually means that a species prefers the aqueous phase over the oil phase. In this example hydrophilic has the same meaning as oleophobic, but such is not always the case.

Hydrophobe   *See* Hydrophobic.

Hydrophobic   A qualitative term referring to the water-avoiding nature of a species (atom, molecule, droplet, particle, etc.). For emulsions hydrophobic usually means that a species prefers the oil phase over the aqueous phase. In this example hydrophobic has the same meaning as oleophilic, but such is not always the case. A functional group of a molecule that is not very water-soluble is referred to as a hydrophobe.

Hydrophobic Bonding   The attraction between hydrophobic species in water that arises from the fact that the solvent–solvent interactions are more favorable than the solvent–solute interactions.

Hydrophobic Effect   The partitioning of a substance from an aqueous phase into (or onto) another phase due to its hydrophobicity. Often characterized by an octanol–water partitioning coefficient. *See also* Solvent-Motivated Sorption.

Hydrophobic Index   An empirical measure of the relative wetting preference of very small solid particles. In one test method, solid particles of narrow size range are placed on the surfaces of a number of samples of water containing increasing concentrations of alcohol (thus providing a range of solvent surface tensions). The percentage alcohol solution at which the particles just begin to become hydrophilic and sink is taken as the hydrophobic index. The corresponding solvent surface-tension value is taken as the critical surface tension of wetting. The technique is also referred to as the film-flotation technique (reference 45) or sink-float method. *See also* Critical Surface Tension of Wetting.

Hydrophobic Interaction   *See* Hydrophobic Bonding.

Hydrophobic Ionogenic Organic Compound (HIOC)   An organic compound that is capable of ionizing, depending upon the solution pH. Upon ionization the properties of the molecule change and its sorption and subsurface migration (in the environment) vary accordingly.

Hydrophobic Organic Contaminant (HOC)   An organic molecule (usually neutral) that has a relatively low solubility in water. Example: many pesticides.

Hydrotrope   Any species that enhances the solubility of another. Example: hydrotropes such as alkyl aryl sulfonates (e.g., toluene sulfonate) are added to detergent formulations to raise the cloud point.

Hyperfiltration   *See* Ultrafiltration.

Ideal Fluid   *See* Inviscid Fluid.

IHP   Inner Helmholtz plane, *see* Helmholtz Double Layer.

ILS   *See* Ionization-Loss Spectroscopy.

Imbibition   The displacement of a nonwetting phase by a wetting phase in a porous medium or a gel; the reverse of drainage.

Immersional Wetting   The process of wetting when a solid (or liquid) that is initially in contact with gas becomes completely covered by an immiscible liquid phase. *See also* Wetting, Spreading Wetting, Adhesional Wetting.

Impingement Separator   *See* Separator.

Inclined-Plate Settling   *See* Lamella Settling.

Increase of Volume upon Foaming   In foaming, 100 times the ratio of gas volume to liquid volume in a foam. Also termed the foaming power.

Indifferent Electrolyte   An electrolyte whose ions have no significant effect on the electric potential of a surface or interface, as opposed to potential-determining ions that have a direct influence on surface charge. This distinction is most valid for low electrolyte concentrations. Example: for the AgI surface in water $NaNO_3$ would be an indifferent electrolyte, but both Ag^+ and I^- would be potential-determining ions.

Induced Gas Flotation   *See* Froth Flotation.

Induction Forces   Debye forces. *See* Dispersion Forces.

Inelastic Scattering   *See* Light Scattering.

Infinite Clusters   *See* Percolation.

Infrared Reflection–Absorption Spectroscopy (IRAS)   A surface vibrational spectroscopic technique for studying adsorbed molecules on crystals. The absorption of infrared radiation due to the adsorbates is determined after reflection from a plane substrate surface. *See also* Table 9.

Inherent Viscosity   In solutions and colloidal dispersions, the natural logarithm of the relative viscosity, all divided by the solute or dispersed-phase concentration. $\eta_{Inh} = C^{-1} \ln(\eta/\eta_o)$. In the limit of vanishing concentration it reduces to the intrinsic viscosity. Also termed the logarithmic viscosity number. *See* Table 4.

Initial Knockdown Capability   *See* Knockdown Capability.

Initial Spreading Coefficient   *See* Spreading Coefficient.

Ink Bottle Pore   A description of one kind of shape of dead-end pore in a porous medium, in which a narrow throat is connected to a larger pore body. *See also* Porous Medium.

Inner Helmholtz Plane (IHP)   *See* Helmholtz Double Layer.

Inner Potential   (1) In the diffuse electric double layer extending outward from a charged interface, the electrical potential at the boundary between the Stern and the diffuse layer is termed the inner electrical potential. Synonyms include the Stern layer potential or Stern potential. *See also* Electric Double Layer, Zeta Potential.

(2) The electric potential inside a phase, in this case termed the Galvani potential. *See also* Galvani Potential, Outer Potential, Jump Potential.

INS   *See* Ion-Neutralization Spectroscopy.

Integral Capacitance of the Electric Double Layer   *See* Capacitance of the Electric Double Layer.

Intercalation   The formation of a layer of one material between layers of another. Certain compounds are capable of expanding clay crystals through intercalation. With salts that are intercalated as the total salt, the process is termed intersalation. Clay minerals containing an intercalation layer are also termed pillar interlayered clay minerals or pillared inorganic layered compounds (PILC).

Interface   The boundary between two immiscible phases, sometimes including a thin layer at the boundary within which the properties of one bulk phase change over to become the properties of the other bulk phase. An interfacial layer of finite specified thickness may be defined. When one of the phases is a gas, the term surface is frequently used.

Interface Emulsion   An emulsion occurring between oil and water phases in a process separation or treatment apparatus. Such emulsions may have a high solids content and are frequently very viscous. In this case the term interface is used in a macroscopic sense and refers to a bulk phase separating two other bulk phases of higher and lower density. Other terms: cuff layer, pad layer, or rag layer emulsions.

Interfacial Film   A thin layer of material positioned between two immiscible phases, usually liquids, whose composition is different from either of the bulk phases.

Interfacial Layer   The layer at an interface that contains adsorbed species. Also termed the surface layer. *See also* Adsorption Space.

Interfacial Potential   *See* Surface Potential.

Interfacial Rheology   *See* Surface Viscosity.

Interfacial Rheometer   *See* Surface Viscometer.

Interfacial Tension   *See* Surface Tension.

Interfacial Viscometer   *See* Surface Viscometer.

Interfacial Viscosity   *See* Surface Viscosity.

Interferometry   An experimental technique in which a beam of light is reflected from a film. Light reflected from the front and back surfaces of the film travels different distances and produces interference phenomena, a study of which allows calculation of the film thickness.

Intermediate Pore   An older term, now replaced by mesopore.

Internal Phase   *See* Dispersed Phase.

Internal Surface   In porous media the surface contained in pores and throats that are in communication with the outside space. *See also* Molecular Sieve Effect. Media having internal porosity also have internal surface area that may be available for sorption reactions. *See also* Activated Carbon.

Intersalation   *See* Intercalation.

Intrinsic Viscosity   The specific viscosity divided by the dispersed-phase concentration in the limits of both the dispersed-phase concentration approaching infinite dilution, and of shear rate approaching zero

$$[\eta] = \lim_{C \to 0} \lim_{\dot{\gamma} \to 0} \eta_{sp}/C$$

Also termed limiting viscosity number. *See* Table 4.

Inverse Micelle   A micelle that is formed in a nonaqueous medium, thus having the surfactants' hydrophilic groups oriented inward away from the surrounding medium.

Inversion   The process by which one type of emulsion is converted to another, as when an O/W emulsion is transformed into a W/O emulsion, and vice versa. Inversion can be accomplished by a wide variety of physical and chemical means.

Invert Emulsion   A water-in-oil emulsion. This is different from the term reverse emulsion used in the petroleum field.

Invert-Oil Mud   An emulsion drilling fluid (mud) of the water-in-oil (W/O) type, and having a high water content. *See also* Oil-Base Mud, Oil Mud.

Inviscid Fluid   An ideal fluid that has no viscosity. Such a fluid cannot support any applied shear stress and flows without any dissipation of energy. Also referred to as an ideal fluid, or a nonviscous fluid.

Ion Exchange   A special kind of adsorption in which the adsorption of an ionic species is accompanied by the simultaneous desorption of an equivalent charge quantity of other ionic species. Ion exchange is commonly used for removing hardness and other metal ions in water treatment. The ion-exchange media can be arranged to provide a specific selectivity. *See also* Sorbent-Motivated Adsorption.

Ionic Strength   A measure of electrolyte concentration given by $I = \frac{1}{2}\sum c_i z_i^2$, where c_i are the concentrations, in moles per liter, of the individual ions, i, and z_i are the ion charge numbers.

Ionization-Loss Spectroscopy (ILS)   A technique related to photoelectron spectroscopy in which the emission of secondary electrons is studied and used for the determination of surface composition. *See also* Table 9.

Ion-Neutralization Spectroscopy (INS)   A surface technique in which low-energy inert gas ions are made to strike a surface and become neutralized by a charge-transfer process that leads to the ejection of electrons, which are detected. Information about both the surface and adsorbed material can be gained. *See also* Table 9.

Ion-Scattering Spectroscopy (ISS)   A scattering technique used for the determination of surface composition by scanning the surface with a monoenergetic ion beam. The energy of the scattered ions is related to the mass of the scattering atoms at the surface so that the masses of the surface atoms can be determined. The techniques employing low-energy ions (less than 10 keV) are termed low-energy ion-scattering spectroscopy (LEIS), or ion-scattering spectroscopy (ISS). The techniques employing high-energy ions (greater than 100 keV) are termed medium-energy ion-scattering (MEIS) or high-energy ion-scattering (HEIS) spectroscopy. *See also* Table 9.

Ion-Selective Membrane   A membrane that is permeability-selective (permselective) for certain ions. Typically such a membrane will carry an electric charge and therefore tend not to be permeable to ions of like charge. Selectivity among ions of opposite charge to the membrane but like charge among each other can sometimes be achieved through adjustment of a membrane's pore sizes.

IRAS   *See* Infrared Reflection–Absorption Spectroscopy.

Irridescent Layers   *See* Schiller Layers.

Isaphroic Lines   Contours of equal foam stability plotted on foam-phase diagrams. Example: *see* reference 32, page 312.

Isobar   The mathematical representation of a phenomenon occurring at constant pressure. *See* Adsorption Isotherm.

Isodisperse   *See* Monodisperse.

Isoelectric   An ionic macromolecule that exhibits no electrophoretic or electro-osmotic motion.

Isoelectric Focusing   A method for the separation of charged colloidal particles or large molecules. An electric field gradient is imposed along a supporting medium as in zone electrophoresis. However, in this case the supporting medium also supports a pH gradient. A sample of mixture to be separated is applied to one end of the supporting medium, and electrophoretic motion of each species occurs until it comes to rest at a pH corresponding to its isoelectric point. Regions of different components separate out along the direction of the electric field and pH gradient according to the different isoelectric points of the components (typically the cathode end is held at the most basic pH). *See also* Zone Electrophoresis.

Isoelectric Point   The solution pH or condition for which the electrokinetic or zeta potential is zero. Under this condition a colloidal system will exhibit no electrophoretic or electro-osmotic motions. *See also* Point of Zero Charge.

Isoionic   An ionic macromolecule system is isoionic if the only other ions in the system are the ions of the solvent, such as H^+ and OH^- in water.

Isoionic Point   The solution pH or condition for which a species has a zero net charge. Under this condition the species may not have zero charge, but rather regions of opposite charges that balance. Example: proteins and kaolinite clay particles may exhibit isoionic points.

Isokinetic Sampling   Collecting samples of a flowing dispersion using a method in which the sampling velocity (in the sampling probe) is equal to the upstream local velocity. If these velocities are not the same (anisokinetic sampling) then fluid streamlines ahead of the probe will be distorted; collection of particles or droplets will be influenced by their inertia, which varies with particle size; and sampling will not be representative.

Isostere   The mathematical representation of a phenomenon occurring at constant volume. *See also* Adsorption Isostere, Adsorption Isotherm.

Isotactic Polymer   *See* Atactic Polymer.

Isotherm   The mathematical representation of a phenomenon occurring at constant temperature. *See also* Adsorption Isotherm.

ISS   *See* Ion-Scattering Spectroscopy.

Jar Test   For emulsions, foams, or water treatment, *see* Bottle Test.

Jet Impingement   A dispersion technique in which a jet of liquid is directed at a surface or at a jet of another liquid.

Jet Mill   A machine for the comminution, or size reduction, of mineral or other particles. Such machines accelerate feed particles in a jet and cause size reduction by promoting interparticle and particle–wall collisions at high speed. Very small-sized particles can be produced with these mills. Also termed jet pulverizers.

Jet Pulverizer   *See* Jet Mill.

Jones–Ray Effect   The decrease in apparent surface tension of water, as determined by capillary rise, due to the addition of small amounts of electrolyte. In reality the surface tension of water increases.

Jump Potential   The difference between the inner (Galvani) potential and the outer (Volta) potential. That is, $1/e$ multiplied by the work required to bring unit charge from just outside a phase into that phase. Also termed surface potential jump or chi potential. See also Inner Potential, Outer Potential.

Kataphoresis   *See* Electrophoresis.

Keesom Forces   *See* Dispersion Forces.

Kelvin Equation   An expression for the vapor pressure of a droplet of liquid, $RT \ln(p/p_o) = 2\gamma V/r$, where p is the vapor pressure of the liquid in bulk, p_o is the vapor pressure of the droplet, γ is the surface tension, V is the molar volume, and r is the radius of the liquid droplet. *See also* Young–Laplace Equation.

Kinematic Viscosity   Kinematic viscosity is the absolute viscosity of a fluid divided by the density. *See* Table 4.

Kinetic Coefficient of Friction   *See* Friction.

Kinetic Stability   Although most colloidal systems are metastable or unstable with respect to the separate bulk phases, they may have an appreciable kinetic stability. That is, the state of dispersion may exist for an appreciable length of time. Colloidal species can come together in very different ways; therefore, kinetic stability can have different meanings. A colloidal dispersion can be kinetically stable with respect to coalescence but unstable with respect to aggregation. Or, a system could be kinetically stable with respect to aggregation but unstable with respect to sedimentation. It is crucial that stability be understood in terms of a clearly defined process. *See also* Colloid Stability, Thermodynamic Stability.

Klevens Constants   The empirical parameters in an equation advanced by Klevens for predicting the critical micelle concentrations (cmc) of surfactants in terms of the number of carbon atoms in the hydrocarbon chain (n): log (cmc) = $A - Bn$, where A and B are the Klevens constants. The Klevens constants for numerous surfactants are tabulated in references 15 and 16.

Knockdown Capability   A measure of the effectiveness of a defoamer. First, a column of foam is generated in a foam stability apparatus and the foam height is recorded. A measured amount of defoamer is added, and the reduction in foam height over a specified time period, for example, 2 s, is noted. The knockdown capability is the reduction in foam height. There are many variations of this test. Sometimes referred to as initial knockdown capability.

Knockout Drops   Demulsifier that may be used to enhance the separation of oil from water and solids, in an emulsion, in the centrifuge test for determining basic sediment and water (BS&W).

Köhler Illumination   In microscopy, the illumination provided through a condenser lens system adjusted to produce optimum brightness with uniform illumination of a sample.

Krafft Point   The temperature above which the solubility of a surfactant increases sharply (micelles begin to be formed). In practice a narrow range of temperatures represents the transition from a solution in which only single, unassociated surfactant molecules (monomers) or ions (ionomers) can be present, up to a given solubility limit, to a solution that can contain micelles and thus allow much more surfactant to remain in solution in preference to precipitating. Numerous tabulations are given in references 15 and 16.

Krafft Temperature   *See* Krafft Point.

Kugelschaum   *See* Gas Emulsion.

Lamella   *See* Foam.

Lamella Division   A mechanism for foam lamella generation in porous media. Typically, when a foam lamella reachs a branch point in a flow channel, the lamella may divide into two lamellae rather than simply follow one of the two available pathways. This is termed lamella division. *See also* Snap-Off, Lamella Leave-Behind.

Lamella Leave-Behind   A mechanism for foam lamella generation in porous media. When gas invades a liquid-saturated region of a porous medium, it may not displace all of the liquid, but rather leave behind liquid lamellae that will be oriented parallel to the direction of the flow. A foam generated entirely by the lamella leave-behind mechanism will be gas-continuous. *See also* Snap-Off, Lamella Division.

Lamella Number   A dimensionless parameter used to predict the liklihood that a combination of capillary suction in plateau borders and the influence of mechanical shear will cause an oil phase to become emulsified and imbibed into foam lamellae flowing in porous media. It predicts that this will happen when $L > 1$, where $L = \Delta P_C/\Delta P_R = (\gamma^o{}_F)r_o/(\gamma_{OF})r_p$. Here ΔP_C and ΔP_R are the pressure differences between inside a plateau border and inside the laminar part of a lamella, and the pressure difference across the oil–aqueous interface respectively; r_o and r_p are the radii of the oil surface with which the lamella comes into contact, and that of the lamella plateau border, respectively; and $\gamma^o{}_F$ and γ_{OF} are the foaming solution surface tension and the foaming solution–oil interfacial tension, respectively. Simplified forms of this equation have also been used (reference 47).

Lamellar Foam   Although all foams contain lamellae, this term is sometimes used to distinguish a certain kind of foam in porous media. When the length scale of the confining space is comparable to or less than the length scale of the foam bubbles, the foam is termed lamellar foam to distinguish from the opposite case, termed bulk foam. *See also* Foam, Foam Texture.

Lamella Settling   A process for phase separation based on density differences. A commercial lamella settler for suspensions or emulsions comprises a stack of parallel plates spaced apart from each other and inclined from the horizontal. The space between each set of plates forms a separate settling zone. The feed is pumped into these spaces, at a point near the longitudinal middle of the plates. The less dense phases rise to the underside of the upper plates and flow to the tops of those plates. Meanwhile, the more dense phases settle down to the upperside of the lower plates and flow to the bottoms of those plates. Product is collected at the tops of the plate stack, and tailings are collected at the bottom of the plate stack. Such an inclined lamella-settling process is much more efficient than vertical gravity separation. Also termed inclined plate settling or inclined tube settling.

Lamina   *See* Foam.

Laminar Flow   A condition of flow in which all elements of a fluid passing a certain point follow the same path, or streamline; there is no turbulence. Also referred to as streamline flow.

Langmuir Adsorption   *See* Adsorption Isotherm.

Langmuir–Blodgett Film   A film of molecules that is deposited onto a solid surface by repeatedly passing the solid through a monolayer of molecules at a gas–liquid interface. Each pass deposits an additional monolayer on the solid. Example: the first applications of Langmuir–Blodgett films were in the preparation of nonglare coatings on glasses and other lenses.

Langmuir Film Balance   *See* Film Balance.

Langmuir Isotherm   An adsorption isotherm equation that assumes monolayer adsorption and constant enthalpy of adsorption. The amount adsorbed per mass of adsorbent is proportional to equilibrium solute concentration at low concentrations and exhibits a plateau or limiting adsorption at high concentrations. *See* Adsorption Isotherm.

Langmuir Trough   *See* Film Balance.

Laplace Flow   *See* Capillary Flow.

Laplace Waves   *See* Capillary Ripples.

Latex   A dispersion (suspension or emulsion) of polymer in water. Latex rubber, a solid, is produced either by coagulating natural latex or by synthetic means through emulsion polymerization. Example: latex paint is a latex containing pigments and filling additives.

Lather   A foam produced by mechanical agitation on a solid surface. Example: the mechanical generation of shaving foam (lather) on a wet bar of soap.

Launderometer   The specialized machine used to perform a standardized test method for measuring the effectiveness of detergents. The degree to which reference soils are washed from standard fabric swatches in the presence of detergents and under specified conditions is determined. *See also* Detergent, Detergency.

Leave-Behind   *See* Lamella Leave-Behind.

LEED   *See* Low-Energy Electron Diffraction.

LEIS   Low-energy ion-scattering spectroscopy. *See* Ion-Scattering Spectroscopy.

Lennard–Jones 6–12 Potential   *See* Lennard-Jones Potential.

Lennard–Jones Potential   A measure of the potential energy of interaction between two atoms. Also termed the Lennard–Jones 6–12 potential.

Lens   Physics: Any piece of material or device that concentrates or disperses an incident beam of light, sound, electrons, or other radiation. Example: the curved pieces of glass that magnify an image formed in a microscope.

Colloid: A nonspreading droplet of liquid at an interface is said to form a lens. The lens is thick enough for its shape to be significantly influenced by gravitational forces.

Geology: A specific geological layer resembling a convex lens. Example: clay mineral lens.

Levitator   An instrument for dielectrophoresis studies in which the motion of dipolar colloidal species is observed in a vertically imposed, nonhomogeneous electric field. Normally, depending on the sign of the Clausius–Mossotti factor, species will tend to move toward either the most intense or the least intense region of the electric field. In this case, by adjusting the frequency of the oscillating electric field (changing the Clausius–Mossotti factor), a state of levitation can be achieved for which the species move neither up nor down. *See also* Dielectrophoresis, Clausius–Mossotti Factor.

Lifshitz–van der Waals Forces   *See* Dispersion Forces.

Light Crude Oil   A naturally occurring hydrocarbon having a viscosity less than 10,000 mPa·s at ambient deposit temperature, and a density less than 934 kg/m^3 at 15.6 °C. *See* references 25–27.

Light Nonaqueous-Phase Liquid (LNAPL)   *See* Nonaqueous-Phase Liquid.

Light Scattering   Light will be scattered (deflected) by local variations in refractive index caused by the presence of dispersed species depending upon their size (reference 4). In elastic scattering there is no wavelength shift, in inelastic scattering there are wavelength shifts due to molecular transitions, and in quasielastic scattering there are wavelength shifts and line broadening due to time-dependent processes. In light scattering the scattering plane contains both the incident light beam and the line that connects the center of the scattering system to the point of observation. The scattering angle lies in this plane and is measured clockwise viewing into the incident beam. By this measure forward scattering is at a scattering angle of zero and backscattering is at a scattering angle of 180°. *See also* Mie Scattering, Nephelometry, Rayleigh Scattering, Tyndall Scattering.

Limiting Capillary Pressure   For foam flow in porous media the maximum capillary pressure that can be attained by simply increasing the fraction of gas flow. Foams flowing at steady-state do so at or near this limiting capillary pressure. In the limiting capillary pressure regime the steady-state saturations remain essentially constant.

Limiting Diffusion Coefficient   *See* Diffusion Coefficient.

Limiting Sedimentation Coefficient   *See* Sedimentation Coefficient.

Limiting Viscosity Number   *See* Intrinsic Viscosity.

Line Tension   Where three phases meet there may exist a line tension (force) along the three-phase junction. For a lens of material at the interface between two other immiscible phases, the three-phase contact junction takes the form of a circle along which the line tension acts.

Lipid   Long-chain aliphatic hydrocarbons and derivatives originating in living cells. Some lipids, such as fatty acids, are also surfactants. Simple lipids tend to be hydrocarbon-soluble but not water-soluble. Examples: fatty acids, fats, waxes.

Lipid Bilayer   *See* Bimolecular Film.

Lipid Film   A thin film of oil in water in which the film is stabilized by lipids. The term is used even though the film is not a film of lipid. *See also* Fluid Film.

Lipophile   That part of a molecule that is organic-liquid-preferring in nature.

Lipophilic   The (usually fatty) organic-liquid-preferring nature of a species. Depending on the circumstances may also be a synonym for oleophilic. *See also* Hydrophile–Lipophile Balance.

Lipophobe   That part of a molecule that is organic-liquid-avoiding in nature.

Lipophobic   The (usually fatty) organic-liquid-avoiding nature of a species. Depending on the circumstances may also be a synonym for oleophobic.

Liposome   *See* Vesicle.

Liquid Aerosol   *See* Aerosol.

Liquid-Crystalline Phase   *See* Mesomorphic Phase.

Liquid Crystals   *See* Mesomorphic Phase.

Liquid Limit   The minimum water content for which a small sample of soil or similar material will barely flow in a standardized test method (references 23 and 24). Also termed the upper plastic limit. *See also* Atterberg Limits, Plastic Limit, Plasticity Number.

LNAPL   Light nonaqueous-phase liquid. *See* Nonaqueous-Phase Liquid.

Logarithmic Viscosity Number   *See* Inherent Viscosity.

London Forces   *See* Dispersion Forces.

Loose Emulsion   A petroleum industry term for a relatively unstable, easy-to-break emulsion, as opposed to a more stable, difficult-to-treat emulsion. *See also* Tight Emulsion.

Lorenz–Mie Scattering   *See* Mie Scattering.

Low-Energy Electron Diffraction (LEED)   A diffraction technique in which a low-energy electron beam is used. In this case the electrons have low penetrating power, and measuring the diffraction pattern yields information about surface structure. The inelastically scattered electrons, having lower energy, are stopped by grids, and the elastically scattered electrons, having the original energy level, are used to form the image pattern. Hence the term elastic low-energy electron diffraction (ELEED) is also used. *See also* High-Energy Electron Diffraction, Table 9.

Lower Plastic Limit   *See* Plastic Limit.

Low-Energy Ion-Scattering Spectroscopy (LEIS)   *See* Ion-Scattering Spectroscopy.

Lower-Phase Microemulsion   A microemulsion that has a high water content and is stable while in contact with a bulk oil phase, and in laboratory tube or bottle tests tends to be situated at the bottom of the tube, underneath the oil phase. For chlorinated organic liquids, which are more dense than water, the oil will be the bottom phase rather than the top. *See* Microemulsion, Winsor-Type Emulsions.

Lubrication   The action of a substance to reduce friction between two materials. Usually the lubricating film is thick enough for the material surfaces to be quite independent of each other.

Boundary lubrication refers to the situation where only a thin film separates the material surfaces and the coefficient of friction depends upon the specific nature of the interfacial region. The science of friction and lubrication is known as tribology.

Lundelius Rule   An expression for the inverse relation between solubility and the extent of adsorption of a species.

Lyocratic   A dispersion stabilized principally by solvation forces. Example: the stability of aqueous biocolloid systems can be explained in terms of hydration and steric stabilization. *See also* Electrocratic.

Lyophilic   General term referring to the continuous-medium- (or solvent)-preferring nature of a species. *See* Hydrophilic.

Lyophilic Colloid   An older term used to refer to single-phase colloidal dispersions. Examples: polymer and micellar solutions. Other synonyms no longer in use: semicolloid or half-colloid.

Lyophobic   General term referring to the continuous-medium- (or solvent)-avoiding nature of a species. *See* Hydrophobic.

Lyophobic Colloid   An older term used to refer to two-phase colloidal dispersions. Examples: suspensions, foams, emulsions.

Lyophobic Mesomorphic Phase   *See* Mesomorphic Phase.

Lyoschizophrenic Surfactant   A surfactant in a two-phase system whose behavior indicates a lack of preference for solubility in one phase or the other (reference 9).

Lyotropic Liquid Crystals   *See* Mesomorphic Phase.

Lyotropic Mesomorphic Phase   *See* Mesomorphic Phase.

Lyotropic Series   A series and order of ions indicating, in decreasing order, their effectiveness in influencing the behavior of colloidal dispersions. Also termed Hofmeister series. Example: the following series shows the effect of different species on coagulating power.

cations: $Cs^+ > Rb^+ > K^+ > Na^+ > Li^+$
anions: $CNS^- > I^- > Br^- > Cl^- > F^- > NO_3^- > ClO_4^-$

Macroemulsion   *See* Emulsion. In enhanced oil recovery terminology the term macroemulsion is employed sometimes to identify
emulsions having droplet sizes greater than some specified value
and sometimes simply to distinguish an emulsion from the
microemulsion or micellar emulsion types.

Macroion   A charged colloidal species whose electric charge is
attributable to the presence at the surface of ionic functionalities.

Macromolecule   A large molecule composed of many simple
units bonded together. Macromolecules may be naturally occurring,
such as humic substances, or synthetic, such as many polymers.

Macropore   *See* Pore.

**Magnetic Resonance Imaging (MRI imaging, nuclear magnet-
ic resonance imaging)**   A technique for imaging and quantify-
ing the distribution of phases in multiphase systems, including dispersions in porous media. The technique employs a homogeneous,
static, high magnetic field with a superimposed, time-dependent,
linear-gradient magnetic field so that the total magnetic- field
strength depends on position in the sample. Resonance is induced
with radiofrequency energy. In the imaging system the position-
dependent resonance frequencies and signal intensities allow the
determination of the concentration, chemical environment, and
position of any NMR-active nuclei in the sample.

Magnetophoretic Mobility   The mobility of a paramagnetic or ferromagnetic particle moving under the influence of an external magnetic field. The magnetophoretic mobility equals the particle velocity, relative to the medium, divided by the magnetic-field gradient at the location of the particle (reference 33). This definition is analogous to the definition of electrophoretic mobility.

Main Active   The primary surfactant in a detergent formulation. *See also* Detergent.

Marangoni Effect   In surfactant-stabilized fluid films, any stretching in the film causes a local decrease in the interfacial concentration of adsorbed surfactant. This decrease causes the local interfacial tension to increase (Gibbs effect), which in turn acts in opposition to the original stretching force. With time the original interfacial concentration of surfactant is restored. The time-dependent restoring force is referred to as the Marangoni effect and is a mechanism for foam and emulsion stabilization. The combination of Gibbs and Marangoni effects is properly referred to as the Gibbs–Marangoni effect, but is frequently referred to simply as the Marangoni effect.

Marangoni Elasticity   *See* Film Elasticity, Marangoni Effect.

Marangoni Flow   Liquid flow in response to a gradient in surface or interfacial tension. *See* Marangoni Effect.

Marangoni Surface Elasticity   *See* Film Elasticity, Marangoni Effect.

Marangoni Waves   *See* Capillary Ripples.

Marine Colloids   Any colloids derived from marine sources. Examples include the hydrophilic colloids (hydrocolloids) derived from various seaweeds, such as algin, and colloids derived from marine animals, such as chitin. *See also* Seaweed Colloids, Chitin.

Mark–Houwink Equation   *See* Staudinger–Mark–Houwink Equation, Table 5.

Martin's Diameter   A statistical particle diameter; the length of a line drawn parallel to a chosen direction such that it bisects the area of a particle. The value obtained depends on the particle orientation, and so these measurments have significance only when a large enough number of measurements are averaged together. *See also* Feret's Diameter.

Mass-Area Mean Diameter   An average particle diameter calculated from measurement of average particle area.

Mass-Average Quantities   A method of averaging in which the sum of the amount of species multiplied by the property of interest squared is divided by the sum of the amount of species multiplied by the property. An example is the mass-average relative molecular mass determined by light-scattering methods,

$$\overline{M}_{r,m} = \frac{\sum n_i M_r^2(i)}{\sum n_i M_r(i)}$$

where n_i is the amount of species and $M_r(i)$ is the relative molecular mass of species i. *See also* Number-Average Quantities.

Maximum Bubble Pressure Method   A method for the determination of surface tension in which bubbles of gas are formed and allowed to dislodge from a capillary tube immersed in a liquid. The maximum bubble pressure achieved during the growth cycle of the bubbles is used to calculate the surface tension on the basis of the pendant-drop analysis method. Variations include the differential maximum bubble pressure method, in which two capillaries are used and the difference in maximum bubble pressures is determined.

Maxwell–Wagner Polarization   A phenomenon in which the ions contained within a cell separate toward opposite sides of, but still within, the cell under the influence of an applied electric field. The charge separation within the cell creates a dipole.

MBS   *See* Molecular Beam Spectroscopy.

Mechanical Impact Mill   A machine for the comminution, or size reduction, of mineral or other particles. Such machines pulverize feed particles (typically about 10 mm initially) by causing them to strike a surface at high speed. Very small-sized particles can be produced with these mills.

Mechanical Syneresis   Any process in which syneresis is enhanced by mechanical means. *See also* Syneresis.

Medium-Energy Ion Scattering (MEIS)   *See* Ion-Scattering Spectroscopy.

Medium Sand   *See* Sand, Table 3.

MEIS   Medium-energy ion-scattering spectroscopy. *See* Ion-Scattering Spectroscopy.

Membrane E.M.F.   *See* Membrane Potential.

Membrane Potential   The potential difference between two identical salt bridges placed into two ionic solutions that are separated from each other by a membrane. *See also* Donnan Equilibrium.

Meniscus   The uppermost surface of a column of a liquid. The meniscus may be either convex or concave depending on the balance of gravitational and surface or interfacial tension forces acting on the liquid.

Mercury Porosimetry   *See* Porosimeter.

Mesomorphic Phase   A phase consisting of anisometric molecules or particles that are aligned in one or two directions but randomly arranged in other directions. Such a phase is also commonly referred to as a liquid-crystalline phase or simply a liquid crystal. The mesomorphic phase is in the nematic state if the molecules are oriented in one direction, and in the smectic state if oriented in two directions. Mesomorphic phases are also sometimes distinguished on the basis of whether their physical properties are mostly determined by interactions with surfactant and solvent (lyotropic liquid crystals) or by temperature (thermotropic liquid crystals). *See also* Neat Soap.

Mesopore   *See* Pore.

Metastable   *See* Thermodynamic Stability.

Micellar Aggregation Number   *See* Aggregation Number.

Micellar Catalysis   Catalytic reactions conducted in a surfactant solution in which micelles play a role in catalyzing the reaction. Typically the micelles either solubilize needed reactant(s) or they provide a medium of intermediate polarity to enhance the rate of a reaction.

Micellar Charge   The net charge of surfactant ions in a micelle including any counterions bound to the micelle.

Micellar Emulsion   An emulsion that forms spontaneously and has extremely small droplet sizes (<10 nm). Such emulsions are thermodynamically stable and are sometimes referred to as microemulsions.

Micellar Mass   The mass of a micelle. For ionic surfactants this value includes the surfactant ions and their counterions.

Micellar Solubilization   *See* Solubilization.

Micellar Weight   *See* Micellar Mass.

Micelle   An aggregate of surfactant molecules or ions in solution. Such aggregates form spontaneously at sufficiently high surfactant concentration, above the critical micelle concentration. The micelles typically contain from tens to hundreds of molecules and are of colloidal dimensions. If more than one kind of surfactant forms the micelles, they are referred to as mixed micelles. If a micelle becomes larger than usual as a result of either the incorporation of solubilized molecules or the formation of a mixed micelle, then the term swollen micelle is applied. *See also* Critical Micelle Concentration, Inverse Micelle.

Microelectrophoresis   *See* Electrophoresis.

Microemulsion   A special kind of stabilized emulsion in which the dispersed droplets are extremely small (<100 nm) and the emulsion is thermodynamically stable. These emulsions are transparent and may form spontaneously. In some usage a lower size limit of about 10 nm is implied in addition to the upper limit; *see also* Micellar Emulsions. In some usage the term microemulsion is reserved for a Winsor type IV system (water, oil, and surfatants all in a single phase). *See also* Winsor-Type Emulsions.

Microencapsulation   The protection of a chemical species by containing it in small droplets, particles, or bubbles covered by a protective coating. Example: the encapsulation of liquid within vesicles.

Microfoam   *See* Gas Emulsion.

Microgas Emulsions   A kind of foam in which the gas bubbles have an unusually thick stabilizing film and exist clustered together as opposed to either separated, nearly spherical bubbles or the more concentrated, system-filling polyhedral bubbles. A microgas emulsion will cream to form a separate phase from water. Also termed aphrons or colloidal gas aphrons.

Micrometer 10^{-6} m, a common distance unit in colloid science. The common symbol is μm; the symbol μ has also been frequently used, but is discouraged by the Systeme Internationale (SI). The millimicron, 10^{-3} μm, has also been used, sometimes abbreviated as mμ, also discouraged. The symbol μμ has sometimes been used where millimicron, mμ, was meant. *See also* Micron.

Micron   In the older literature micron was one of three particle size range distinctions that were judged on the basis of visibility under the dark-field or bright-field microscope. Particles visible under the bright-field microscope were termed microns (diameters greater than about 500 nm). Particles not visible under the bright-field microscope but visible under the dark-field microscope (ultramicroscope) were termed submicrons. Particles that were not even visible under the dark-field microscope were termed amicrons (diameters less than about 5–50 nm). These distinctions are no longer in use. The term micron to indicate 10^{-6} is discouraged by the Systeme Internationale; the correct term is micrometer. *See also* Micrometer.

Micronizing   The process by which a solid is reduced to particle sizes of less than about 100 μm, using any type of particle size reduction equipment. Examples: micronized talc, micronized pigment.

Microphotograph   A photographic image forming a small copy of a much larger object. This image is not the same as a photomicrograph.

Micropore   *See* Pore.

Micropore Filling   The process by which molecules become adsorbed within micropores.

Micropore Volume   The volume of adsorbed material that completely fills the micropores in a porous medium, expressed in terms of liquid volume at atmospheric pressure and specified temperature.

Microscopic Electrophoresis   *See* Electrophoresis.

Microscopy   Light microscopy involves the use of light rays and lenses to observe magnified images of objects. The magnified image may be formed from transmitted light for transparent materials, or from reflected (incident) light for opaque materials. In each case there are different illuminating modes, and the light used may be visible, infrared, or ultraviolet. *See also* Bright-Field Illumination, Dark-Field Illumination, Köhler Illumination. Different viewing modes may be used, such as polarizing, fluorescence, phase contrast, and interference contrast. A derived technique is confocal microscopy. An analogous technique, electron microscopy, involves the use of electrons rather than light. *See also* Confocal Microscopy, Electron Microscopy.

Microtome Method   A means of determining the surface concentration of species by cutting away a thin surface layer with a knife (microtome), physically separating the layer, and analyzing it.

Middle-Phase Microemulsion   A microemulsion that has high oil and water contents and is stable while in contact with either bulk oil or bulk water phases. This stability may be due to a bicontinuous structure in which both oil and water phases are continuous at the same time. In laboratory tube or bottle tests involving samples containing unemulsified oil and water, a middle-phase microemulsion will tend to be situated between the two phases. *See also* Bicontinuous Microemulsion, Winsor-Type Emulsions.

Middle Soap   *See* Neat Soap.

Mie Scattering   Light will be scattered (deflected) by local variations in refractive index caused by the presence of dispersed species, depending upon their size. The scattering of light by species whose size is much less than the wavelength of the incident light is referred to as Rayleigh scattering, and it is termed Mie scattering if the species' size is comparable to that of the incident light. Also termed Lorenz–Mie scattering. *See also* Lorenz–Mie Scattering, Light Scattering, Rayleigh Scattering.

Millimicron   *See* Micron.

Miniemulsion   *See* Emulsion. The term miniemulsion is sometimes used to distinguish an emulsion from the microemulsion or micellar emulsion types. Thus a miniemulsion would contain droplet sizes greater than 100 nm and less than 1000 nm or some other specified upper size limit.

Mist   A dispersion of a liquid in a gas (aerosol of liquid droplets) in which the droplets have diameters less than a specified size. In industry mists have droplet sizes of less than 10 μm, as opposed to sprays, in which the droplet sizes are greater. In the atmosphere mists are aerosols of liquid droplets having sizes similar to those given for clouds in Table 1.

Mitchell Foam Quality   *See* Foam Quality.

Mixed Micelles   *See* Micelle.

Mobile Film   *See* Fluid Film.

Mobility Reduction Factor (MRF)   A dimensionless measure of the effectiveness of a foam at reducing gas mobility when flowing in porous media. In one definition, the mobility reduction factor is equal to the mobility (or pressure drop) measured for foam flowing through porous media divided by the mobility (or pressure drop) measured for surfactant-free solution and gas flowing at the same volumetric flow rates.

Modulus of Surface Elasticity   *See* Film Elasticity.

Molecular Beam Spectroscopy (MBS)   A technique for studying the kinetics of reactions at surfaces, in which a molecular beam strikes a surface and the lag time before which reaction products appear is determined.

Molecular Sieve Effect   In porous media the amount of internal surface accessible to molecules may depend on the size of the molecules and may be different for various components in a mixture. The different extent of internal surface experienced by different molecules is termed the molecular sieve effect. *See also* Internal Surface.

Monodisperse   A colloidal dispersion in which all the dispersed species (droplets, particles, etc.) have the same size. Otherwise, the system is heterodisperse (paucidisperse or polydisperse).

Monolayer Adsorption   Adsorption in which a first or only layer of molecules becomes adsorbed at an interface. In monolayer adsorption, all of the adsorbed molecules will be in contact with the surface of the adsorbent. The adsorbed layer is termed a monolayer or monomolecular film.

Monolayer Capacity   In chemisorption, the amount of adsorbate needed to satisfy all available adsorption sites. For physisorption, the amount of adsorbate needed to cover the surface of the adsorbent with a complete monolayer.

Monomolecular Film   *See* Monolayer Adsorption.

Monomolecular Layer   *See* Monolayer Adsorption.

Monopolar   A polar substance that has only one kind of polar properties, either electron-donor or electron-acceptor.

Mooney Equation   An empirical equation for estimating the viscosity of an emulsion. *See* Table 5.

Motionless Mixer   *See* Static Mixer.

Mousse Emulsion   *See* Chocolate Mousse Emulsion.

Moving Boundary Electrophoresis   An indirect electrophoresis technique for particles too small to be viewed. This principle is used in the Tiselius apparatus. Here a colloidal dispersion is placed in the bottom of a U-tube, the upper arms of which are filled with a less dense liquid that both provides the boundaries and makes the connections to the electrodes. Under an applied electric field the motions of the ascending and descending boundaries are measured.

MRF   *See* Mobility Reduction Factor.

MRI Imaging   *See* Magnetic Resonance Imaging.

Multilayer Adsorption   Adsorption in which the adsorption space contains more than a single layer of molecules; therefore, not all adsorbed molecules will be in contact with the surface of the adsorbent. *See also* Brunauer–Emmett–Teller Isotherm, Monolayer Adsorption.

Multiple Emulsion   An emulsion in which the dispersed droplets themselves contain even more finely dispersed droplets of a separate phase. Thus, there may occur oil-dispersed-in-water-dispersed-in-oil (O/W/O) and water-dispersed-in-oil-dispersed-in-water

(W/O/W) multiple emulsions. More complicated multiple emulsions such as O/W/O/W and W/O/W/O are also possible.

Myelin Cylinders   Long-chain polar compounds, above their solubility limit, may interact with surfactants to form mixed micelles that separate (as a coacervate) in the form of cylinders. These are termed myelin cylinders or myelinic figures. They are usually quite viscous and may be birefringent.

Myelinic Figures   *See* Myelin Cylinders.

Naphtha   A petroleum fraction that is operationally defined in terms of the distillation process by which it is separated. A given naphtha is thus defined by a specific range of boiling points of its components. Naphtha is sometimes used as a diluent for W/O emulsions.

NAPL   *See* Nonaqueous-Phase Liquid.

Neat Phase   *See* Neat Soap.

Neat Soap   A mesomorphic (liquid-crystal) phase of soap micelles, oriented in a lamellar structure. Neat soap contains more soap than water. Neat soap is in contrast to middle soap, which contains less soap than water and is also a mesomorphic phase, but has a hexagonal array of cylinders rather than a lamellar structure. *See also* reference 4.

Negative Adsorption   *See* Adsorption.

Negative Tactoids   *See* Tactoids.

Nelson-Type Emulsions   Several types of phase behavior occur in microemulsions; they are denoted as Nelson type II–, type II+, and type III. These designations refer to equilibrium phase behaviors and distinguish, for example, the number of phases that may be in equilibrium and the nature of the continuous phase. *See also* reference 34. Winsor-type emulsions are similarly identified, but with different type numbers.

Nematic State   *See* Mesomorphic Phase.

Nephelometric Turbidity Unit (NTU)   A unit of measurement in nephelometry (turbidity).

Nephelometry   The study of the light-scattering properties of dispersions. In general, a nephelometer is an instrument capable of measuring light scattering by dispersions at various angles. *See also* Light Scattering, Turbidity.

Neumann's Triangle   At the junction where three phases meet, three vectors representing the forces of interfacial tension among pairs of phases can be drawn. At equilibrium, the sum of these vectors of Neumann's triangle will equal zero.

Neutral Agarose   One of the kinds of polysaccharide structure comprising agar. Also termed agaran. *See also* Agar.

Newton Black Film   *See* Black Film.

Newtonian Flow   Fluid flow that obeys Newton's law of viscosity. Non-Newtonian fluids may exhibit Newtonian flow in certain shear-rate or shear-stress regimes. *See also* Newtonian Fluid.

Newtonian Fluid   A fluid or dispersion whose rheological behavior is described by Newton's law of viscosity. Here shear stress is set proportional to shear rate. The proportionality constant is the coefficient of viscosity, or simply, viscosity. The viscosity of a Newtonian fluid is a constant for all shear rates.

Nitrified Foam   A slang term used in some industries to denote foams in which nitrogen is the gas phase.

NMR Imaging   *See* Magnetic Resonance Imaging.

Nomarski Microscopy   A kind of reflected-light microscopy in which a differential interference contrast technique is used to render a relief-like image in interference colors.

Nonaqueous-Phase Liquid (NAPL)   Any liquid other than water. In environmental fields this commonly refers to petroleum hydrocarbons less dense than water (light nonaqueous-phase liquid, LNAPL), or oils such as chlorinated hydrocarbons that are more dense than water (dense nonaqueous-phase liquid, DNAPL).

Nondraining Polymer   Polymer molecules for which the interior of the coiled portions of the molecules are not affected by flow.

Nonionic Surfactant   A surfactant molecule whose polar group is not electrically charged. Example: poly(oxyethylene) alcohol, $C_nH_{2n+1}(OCH_2CH_2)_mOH$.

Non-Newtonian Flow   Fluid flow that does not obey Newton's law of viscosity. Non-Newtonian fluids may exhibit non-Newtonian flow only in certain shear-rate or shear-stress regimes. A number of categories of non-Newtonian flow are distinguished, including dilatant, pseudoplastic, thixotropic, rheopectic, and rheomalaxic. *See also* Newtonian Fluid.

Non-Newtonian Fluid   A fluid whose viscosity varies with applied shear rate (flow rate). *See* Newtonian Fluid.

Nonviscous Fluid   *See* Inviscid Fluid.

Nonwetting   *See* Wetting.

Normal Photoelectron Diffraction (NPD)   *See* Photoelectron Diffraction.

NPD   Normal photoelectron diffraction. *See* Photoelectron Diffraction.

NTU   *See* Nephelometric Turbidity Unit.

Nuclear Magnetic Resonance Imaging   *See* Magnetic Resonance Imaging.

Nuclei   As a solute becomes insoluble, the formation of a new phase has its origin in the formation of clusters of solute molecules, termed germs, that increase in size to form small crystals or particles, termed nuclei. One means of preparing colloidal dispersions involves precipitation from solution onto nuclei, which may be of the same or different chemical species. *See also* Condensation Methods.

Number-Average Quantities   A method of averaging in which the sum of the amount of species multiplied by the property of interest is divided by the total amount of species. An example is the number- average relative molecular mass determined by osmotic pressure measurements,

$$\overline{M}_{r,m} = \frac{\Sigma n_i M_r(i)}{\Sigma n_i}$$

where n_i is the amount of species and $M_r(i)$ is the relative molecular mass of species i. *See also* Mass-Average Quantities.

Numerical Aperture   An indication of the ability of a lens to gather and transmit light.

Oakes Mixer   A machine used for preparing foams in the food industry. A slurry is continuously stirred and aerated under pressure between a series of blades. *See also* Aerator.

Octanol–Water Partition Coefficient   The partitioning coefficient of a compound between octanol and water, that is, between specific nonpolar and polar phases. Used as an indication of the tendency of a compound to partition between oil and water phases. A variety of empirical equations estimate such partitioning of a compound on the basis of its octanol–water partition coefficient. *See also* Solvent-Motivated Sorption.

Oden's Balance   An apparatus for determining sedimentation rates in which a balance pan is immersed in a sedimentation column and is used to intercept and accumulate sedimenting particles, whose mass can be determined as a function of time.

Ohnesorge Equation   An expression giving the critical velocity, V_0, needed for a liquid jet to break up into droplets and form an emulsion, as: $\eta/(\rho\gamma D)^{1/2} = 2000(\eta/[V_0\rho D])^{4/3}$, where η and ρ are the viscosity and density of the liquid in the jet, respectively; γ is the interfacial tension; and D is the nozzle diameter. It has been suggested (reference 35) that this equation should instead be referred to as the Richardson equation.

OHP   Outer Helmholtz plane. *See* Helmholtz Double Layer.

Oil   Liquid petroleum (sometimes including dissolved gas) that is produced from a well. In this sense oil is equivalent to crude oil. The term oil is, however, frequently more broadly used and may include, for example, synthetic hydrocarbon liquids, bitumen from oil (tar) sands, fractions obtained from crude oil, and liquid fats (e.g., triglycerides). *See also* Crude Oil.

Oil-Base Mud   An emulsion drilling fluid (mud) of the water-dispersed-in-oil (W/O) type having a low water content. *See also* Oil Mud, Invert-Oil Mud.

Oil Color   A qualitative test for the presence of emulsified water in an oil. Emulsified water droplets tend to impart a hazy appearance to the oil.

Oil-Emulsion Mud   An emulsion drilling fluid (mud) of the oil-dispersed-in-water (O/W) type. *See also* Oil Mud.

Oil Hydrosol   An oil-in-water (O/W) emulsion in which the oil droplets are very small and the volume fraction of oil is also very small. The emulsion terminology is preferable.

Oil Mud   An emulsion drilling fluid (mud) of the water-dispersed-in-oil (W/O) type. A mud of low water content is referred to as an oil-base mud, and a mud of high water content is referred to as an invert-oil mud. *See also* Oil-Emulsion Mud.

Oleophilic   The oil-preferring nature of a species. A synonym for lipophilic. *See also* Hydrophobic.

Oleophobic   The oil-avoiding nature of a species. A synonym for lipophobic. *See also* Hydrophilic.

Oligomer   A relatively short-chain polymer, typically having a degree of polymerization of less than about 10.

Oliver–Ward Equation   An empirical equation for estimating the viscosity of a dispersion. *See* Table 5.

O/O   Abbreviation for an oil-dispersed-in-oil emulsion in which one oil is polar and the other is not. Example: an emulsion of ethylene glycol in a liquid alkane.

Opacifiers   Agents that make a liquid appear more opaque, or pearlescent. For example, polystyrene latex is added to liquid detergents formulated for dishwashing or shampooing to give them a flat opaque appearance. *See also* Detergent.

Optical Brighteners   Agents that make treated materials appear more white. For example, fluorescent whitening agents are added to laundry detergents so that they may become attached to fibers and give an enhanced whiteness by absorbing UV light and emitting blue light. *See also* Detergent.

Optically Stimulated Exoelectron Emission Spectroscopy (OSEE)   A means of determining the presence and nature of adsorbed species by measuring the electron emission caused by having a beam of light strike a surface in the presence of an electric potential field. *See also* Table 9.

Optimum Salinity   In microemulsions, the salinity for which the mixing of oil with a surfactant solution produces a middle-phase microemulsion containing an oil-to-water ratio of 1. In micellar enhanced oil recovery processes, extremely low interfacial tensions result, and oil recovery tends to be maximized when this condition is satisfied.

Orientation Forces   Keesom forces. *See* Dispersion Forces.

Oriented-Wedge Theory   An empirical generalization used to predict which phase in an emulsion will be continuous and which dispersed. It is based on a physical picture in which emulsifiers are considered to have a wedge shape and will favor adsorbing at an interface such that most efficient packing is obtained, that is, with the narrow ends pointed toward the centers of the droplets. A useful starting point, but there are many exceptions. *See also* Bancroft's Rule, Hydrophile–Lipophile Balance.

Orthokinetic Aggregation   The process of aggregation induced by hydrodynamic motions such as stirring, sedimentation, or convection. Orthokinetic aggregation is distinguished from perikinetic aggregation, the latter being caused by Brownian motions.

Oscillating Jet Method   A method for the determination of surface or interfacial tension in which a stream of gas or liquid is injected into another liquid phase through a jet having an elliptical orifice. Oscillations develop in the jet. The jet dimensions and wavelength are measured and used to calculate the surface or interfacial tension. Also termed the elliptical jet method.

OSEE   *See* Optically Stimulated Exoelectron Emission Spectroscopy.

Osmometer   An instrument for determining the osmotic pressure exerted by solvent molecules diffusing through a semipermeable membrane in contact with a solution or hydrophilic colloidal dispersion. *See also* Colloid Osmotic Pressure, Osmotic Pressure.

Osmosis   The process in which solvent will flow through a semipermeable membrane (permeable to solvent but not to solute) from a solution of lower dissolved solute activity (concentration) to that of higher activity (concentration). *See also* Donnan Equilibrium, Osmotic Pressure.

Osmotic Pressure   When a solution of dissolved species is separated from pure solvent by a semipermeable membrane, not permeable to the dissolved species, the osmotic pressure is the pressure difference required to prevent transfer of the solvent, that is, to prevent osmosis. *See also* Colloid Osmotic Pressure.

Ostwald Ripening   The process by which larger droplets or particles grow in size in preference to smaller droplets or particles because of their different chemical potentials. *See also* Aging.

Ostwald Viscometer   *See* Capillary Viscometer.

Outer Helmholtz Plane (OHP)   *See* Helmholtz Double Layer.

Outer Potential   The potential just outside the interface bounding a specified phase. Also termed the Volta potential. The difference in outer (Volta) potentials between two phases in contact is equal to the surface or interfacial potential between them. *See also* Inner Potential, Jump Potential.

Outgassing   The desorption of gas from a solid under conditions of extremely high vacuum and temperature.

Overpotential   The additional electrical potential beyond that of the thermodynamic electrode potential required to cause current to flow in an electrochemical cell. Also termed overvoltage.

Overvoltage   *See* Overpotential.

O/W   Abbreviation for an oil-dispersed-in-water emulsion.

O/W/A   Abbreviation for a fluid film of water between oil and air. *See* Fluid Film.

O/W/O   In multiple emulsions: Abbreviation for an oil-dispersed-in-water-dispersed-in-oil multiple emulsion. The water droplets have oil droplets dispersed within them, and the water droplets themselves are dispersed in oil forming the continuous phase.

In fluid films: Abbreviation for a thin fluid film of water in an oil phase. Not to be confused with the multiple emulsion convention. *See also* Fluid Film.

Pad Layer Emulsion   *See* Interface Emulsion.

Palisade Layer   In a micelle, the region of water molecules of hydration postulated to lie between relatively water-free hydrocarbon chains at the center of the micelle and the exposed, fully hydrated polar groups at the micelle surface.

Pallmann Effect   *See* Suspension Effect.

Parachor   An empirical parameter used in the estimation of the surface tension of liquids. The parachor, $P = M \gamma^{1/4}/\Delta\rho$, where γ is the surface or interfacial tension, $\Delta\rho$ is the density difference between the phases, and M is the molecular mass of the liquid. *See* Table 8.

Paramagnetic   A material that acquires magnetic properties when placed in an external magnetic field and becomes attracted into the magnetic field (the attraction is proportional to the strength of the applied field). In contrast, diamagnetic materials are weakly repelled by an external magnetic field, and ferromagnetic materials have magnetic properties independently of external magnetic fields.

Particle Size Classification   The separation or determination of particles into different size ranges. A number of classification systems are used, some of which correspond to physical means of separations such as by sieves. Examples are given in Table 3.

PAS   *See* Photoacoustic Spectroscopy.

Passivation   The process by which a nonconducting crystal layer is caused to grow on the surface of a conductor or semiconductor surface to protect the surface from electrical conduction and chemical reaction (e.g., corrosion).

Paucidisperse   A colloidal dispersion in which the dispersed species (droplets, particles, etc.) have a few different sizes. Paucidisperse is a category of heterodisperse systems. *See also* Monodisperse.

PCS   *See* Photon Correlation Spectroscopy.

Peel Test   A means of determining the strength of a joint between two materials. The force normal to the surface required to separate the joint is measured (adhesion), or alternatively, the force parallel to the surface needed to separate the joint (shear adhesion) is measured.

Pendant Bubble Method   *See* Pendant-Drop Method.

Pendant-Drop Method   A method for determining surface or interfacial tension based on measuring the shape of a droplet hanging from the tip of a capillary (in interfacial tension the droplet may alternatively hang upwards from the tip of an inverted capillary). Also termed the hanging drop (or bubble) method.

Peptization   The dispersion of an aggregated (coagulated or flocculated) system. Deflocculation means the same thing.

Peptizing Ions   *See* Potential-Determining Ions.

Percolation   A condition in a dispersed system in which a property such as conductivity increases strongly at a critical concentration, termed the percolation threshold, as a result of the formation of continuous conducting paths, termed infinite clusters.

Percolation Threshold   *See* Percolation.

Perikinetic Aggregation   The process of aggregation when induced by Brownian motions. Perikinetic aggregation is distinguished from orthokinetic aggregation, the latter being caused by hydrodynamic motions such as sedimentation or convection.

Permeability   A measure of the ease with which a fluid can flow (fluid conductivity) through a porous medium. Permeability is defined by Darcy's law. For linear, horizontal, isothermal flow permeability is the constant of proportionality between flow rate times viscosity and the product of cross-sectional area of the medium and pressure gradient along the medium.

Permeate   See Dialysis.

Permittivity   A measure of the ability of a medium to affect an applied electric field. It is the constant of proportionality between the force acting between two point charges and the product of the two charges divided by the square of the distance separating them. The relative permittivity of a material equals the permittivity of the material multiplied by that of vacuum. An older term for relative permittivity is dielectric constant. See also Table 7.

Perrin Black Film   A Newton black film. See Black Film.

Persorption   Selective adsorption of small molecules rather than large molecules in the pores of an adsorbent due to the small sizes of the pores. For such an adsorbent the amount of adsorption varies with the molecular size of the adsorbate. See also Molecular Sieve Effect.

PES   See Photoelectron Spectroscopy.

Petroleum   A general term that may refer to any hydrocarbons or hydrocarbon mixtures, usually liquid, but sometimes solid or gaseous.

Phase Diagram   A graphical representation of the equilibrium relationships between phases in a system. For multicomponent systems, and considering varying temperatures, more than a simple two-dimensional phase diagram will be required.

Phase Inversion Temperature (PIT)   The temperature at which the hydrophilic and oleophilic natures of a surfactant are in balance. As temperature is increased through the PIT, a surfactant will change from promoting one kind of emulsion, such as O/W, to another, such as W/O. Also termed the HLB temperature.

Phase Map   See Phase Diagram.

Phase Ratio   In emulsions phase ratio refers to the ratio between internal phase and continuous phase. Phase ratios are dimensionless, but the units used should be specified because mass ratios and volume ratios are commonly used.

Phase Rule   The fundamental thermodynamic equation governing the equilibria between phases in a system. Also termed the Gibbs phase rule, it specifies the number of intensive variables needed to describe a system (degrees of freedom, f) in terms of the numbers of components, c, and phases, p, present as $f = c - p + 2$.

Phase-Transfer Catalysis   A catalytic reaction in which a surfactant is used to form an ion-pair with a water-soluble reactant and aid the transport of the reactant into an organic phase, where it reacts with other reactant(s) soluble only in the organic phase. *See also* Micellar Catalysis.

PhD   *See* Photoelectron Diffraction.

Phospholipid   Esters of phosphoric acid that contain fatty acid(s), an alcohol, and a nitrogen-containing base. *See also* Lipid.

Phospholipid Bilayer   *See* Bimolecular Film.

Photoacoustic Spectroscopy (PAS)   A technique for studying the vibrational states of surface species and species adsorbed on surfaces. A sample is placed inside a gas-containing chamber and irradiated with radiation of a given wavelength having an intensity modulated at an acoustic frequency. Absorption of this modulated radiation causes periodic heat flow from the sample and generates sound waves that are detected. *See also* Table 9.

Photoelectron Diffraction (PhD)   A diffraction technique similar to high-energy electron diffraction except that a higher energy (X-ray) beam is used at shallow angles to cause the ejection of photoelectrons. Again, the diffraction pattern yields information about surface structure. The distinctions normal photoelectron diffraction (NPD) and azimuthal photoelectron diffraction (APD) are sometimes used. *See also* Table 9.

Photoelectron Spectroscopy (PES)   A technique related to Auger electron spectroscopy and also used for the determination of surface composition. The surface is scanned with a photon beam causing the ejection of electrons from the surface atoms. The energies of the ejected electrons are determined (both primary photoelectrons and the Auger electrons); these energies are characteristic of the atoms from which they were ejected. The principal types of photoelectron spectroscopy use different photon sources: ultraviolet photoelectron spectroscopy (UPS) and X-ray photoelectron spectroscopy (XPS). When XPS is used primarily for chemical analysis, the technique has been referred to as electron spectroscopy for chemical analysis (ESCA), although the term XPS is mostly used now because other electron spectroscopic techniques can also be used for chemical information. *See also* Auger Electron Spectroscopy, Table 9.

Photographic Emulsion   Not an emulsion but rather a dispersion (solid suspension) of silver halide particles in gelatin.

Photomicrograph   A photographic image formed by a microscope. The resulting photographic image is much larger than the original object being photographed. This is not the same as a microphotograph.

Photon Correlation Spectroscopy (PCS)   A means of determining particle size; scattered photons are correlated with the microscopic motion of particles suspended in a fluid. The time fluctuation of the scattering is related to the diffusion of the particles in the medium, and therefore to their size. Also termed quasi-elastic light scattering (QELS).

Photon-Stimulated Desorption Spectroscopy (PSD)   A surface technique in which electronically stimulated adsorbed species are desorbed from a surface to gain information about adsorbate–substrate bonding and about surface composition. In photon-stimulated desorption, photon bombardment is used to excite the adsorbed species, whereas in electron-stimulated desorption (ESD), low-energy electrons (less than 500 eV) are used. *See also* Temperature-Programmed Reaction Spectroscopy (TPRS).

Physical Adsorption   *See* Chemisorption.

Physisorption   *See* Chemisorption.

Pickering Emulsion   An emulsion stabilized by fine particles. The particles form a close-packed structure at the oil–water interface, with significant mechanical strength, which provides a barrier to coalescence.

Pigment   Insoluble material that is finely divided, micronized (for example), and uniformly dispersed in a formulated system for the purpose of coloring it or making it opaque. Examples: TiO_2 in soap bars and paints; iron oxides in eye makeup and paints.

Pigment Grind   Pigment particles dispersed in a liquid, such as castor oil. *See also* Roll Mill.

PILC   *See* Pillar Interlayered Clay Minerals.

Pillar Interlayered Clay Minerals   Clay mineral particles frequently carry a significant electrical charge, which is compensated for by counterions. If the counterions are very large then those present between clay layers cause a significant interlayer separation. Such materials containing an intercalation layer are termed pillar interlayered clay minerals or pillared inorganic layered compounds (PILC). *See also* Intercalation.

PIT   *See* Phase Inversion Temperature.

Plait Point   In phase diagrams the composition condition for which three coexisting phases, containing partially soluble components, of a three-phase system all approach the same composition.

Plasmalemma   *See* Cell Membrane.

Plasma Membrane   *See* Cell Membrane.

Plastic Flow   The deformation or flow of a solid under the influence of an applied shear stress.

Plastic Fluid   A fluid characterized by both of the following: the existence of a finite shear stress that must be applied before flow begins (yield stress), and Newtonian flow at higher shear stresses. May be referred to as Bingham plastic. *See also* Generalized Plastic Fluid.

Plasticity Index   *See* Plasticity Number.

Plasticity Number   The difference between the liquid limit and the plasticity limit of a soil or similar material (references 23 and 24). Also termed the plasticity index. *See also* Atterberg Limits, Liquid Limit, Plastic Limit.

Plastic Limit   The minimum water content for which a small sample of soil or similar material will barely deform or crumble in a standardized test method (references 23 and 24). Also termed the lower plastic limit. *See also* Atterberg Limits, Liquid Limit, Plasticity Number.

Plateau Border   The region of transition at which thin fluid films are connected to other thin films or mechanical supports such as solid surfaces. For example, in foams plateau borders form the regions of liquid situated at the junction of liquid lamellae. Sometimes referred to as a Gibbs ring or Gibbs–Plateau Border.

PLAWM Trough   Pockels–Langmuir–Adam–Wilson–McBain trough. *See* Film Balance.

Pockels–Langmuir–Adam–Wilson–McBain Trough   *See* Film Balance.

Pockels Point   When surfactant molecules are added into a system and form an insoluble film at an interface, surface tension does not decrease very strongly until enough is added to form a complete monolayer. The transition point is termed the Pockels point and corresponds to a surface area occupied per molecule of about 20 $Å^2$ for soaps.

Point of Zero Charge   The condition, usually the solution pH, at which a particle or interface is electrically neutral. This is not always the same as the isoelectric point, which refers to zero charge at the shear plane that exists a small distance away from the interface. *See also* Electrocapillarity, Colloid Titration.

Poiseuille's Law   Poiseuille flow is the steady flow of incompressible fluid parallel to the axis of a circular pipe or capillary. Poiseuille's law is an expression for the flow rate of a liquid in such tubes. It forms the basis for the measurement of viscosities by capillary viscometry.

Poiseuille Flow   *See* Poiseuille's Law.

Poison   In catalysis, any substance that interacts with a catalyst and thereby causes a reduction in catalytic activity, even when present in small concentration. Example: trace sulfur can poison platinum-based catalysts.

Poisson–Boltzmann Equation   A fundamental equation describing the distribution of electric potential around a charged species or surface. The local variation in electric-field strength at any distance from the surface is given by the Poisson equation, and the local concentration of ions corresponding to the electric-field strength at each position in an electric double layer is given by the Boltzmann equation. The Poisson–Boltzmann equation can be combined with Debye–Hückel theory to yield a simplified, and much used, relation between potential and distance into the diffuse double layer.

Poisson Equation   A fundamental equation describing the reduction in electric-field strength that occurs with increasing distance away from a charged species in a dielectric medium. In electric-double-layer theory the effects of the various ion charges are averaged into layers by assuming charge distribution to be a continuous function of distance away from a charged surface. The Poisson equation gives the relationship between the volume charge density at a point in solution and the potential at that same point. This equation can be combined with the Boltzmann equation and Debye–Hückel theory to yield a simplified, and much used, relation between potential and distance into the diffuse double layer.

Polanyi Isotherm   An adsorption isotherm equation that allows for multilayer adsorption and treats adsorbing species as falling into a potential energy minimum. The adsorbed layer is most compressed near the surface and becomes progressively less dense with distance away from the surface. *See* Adsorption Isotherm.

Polar Group   *See* Head Group.

Polar Substance   A substance having different, usually opposite, characteristics at two locations within it. Example: a permanent dipole. Increasing polarity generally increases solubility in water.

Polyacid   *See* Polyelectrolyte.

Polyampholyte   *See* Polyelectrolyte.

Polyanion   *See* Polyelectrolyte.

Polybase   *See* Polyelectrolyte.

Polycation   *See* Polyelectrolyte.

Polydisperse   A colloidal dispersion in which the dispersed species (droplets, particles, etc.) have a wide range of sizes. Polydisperse is a category of heterodisperse systems. *See also* Monodisperse.

Polyederschaum   *See* Foam.

Polyelectrolyte   A kind of colloidal electrolyte comprising a macromolecule that, when dissolved, dissociates to yield a polyionic parent macromolecule and its corresponding counterions. Also termed a polyion, polycation, or polyanion. Similarly, a polyelectrolyte may be referred to in certain circumstances as a polyacid, polybase, polysalt, or polyampholyte (reference 4). Example: carboxymethylcellulose.

Polyion   *See* Polyelectrolyte.

Polymer   A molecule that is made up of many repeating units, or groups, of atoms. Sometimes termed homopolymer to distinguish from copolymers such as block copolymers. *See also* Associative Polymer, Block Copolymer, Atactic Polymer.

Polysalt   *See* Polyelectrolyte.

Polywater   A once-postulated form of water having unusually high viscosity, high surface tension, and low vapor pressure, among other properties. Also termed anomalous water, water II, and cyclimetric water. By 1974 it was determined that polywater does *not* exist, and that the original experimental observations were attributable to other causes. *See* the review by Adamson (reference 10).

Pore   In porous media, the interconnecting channels forming a continuous passage through the medium are made up of pores, or openings, which may be of different sizes. Macropores have diameters greater than about 50 nm. Mesopores have diameters of between about 2.0 and 50 nm. Micropores have diameters of less than about 2.0 nm.

Porosimeter   An instrument for the determination of pore size distribution by measuring the pressure needed to force liquid into a porous medium and applying the Young–Laplace equation. If the surface tension and contact angle appropriate to the injected liquid are known, pore dimensions can be calculated. A common liquid for this purpose is mercury, hence the term mercury porosimetry.

Porosity   The ratio of the volume of all void spaces to total volume in a porous medium. In geology primary porosity refers to initial, or unweathered, media, and secondary porosity refers to that associated with weathered media.

Porous Medium   A solid containing voids or pore spaces. Normally such pores are quite small compared to the size of the solid and well-distributed throughout the solid. In geologic formations porosity may be associated with unconsolidated (uncemented) materials, such as sand, or a consolidated material, such as sandstone.

Potential-Determining Ions   Ions whose equilibrium between two phases, frequently between an aqueous solution and a surface or interface, determines the difference in electrical potential between the phases, or at the surface. Example: for the AgI surface in water both Ag^+ and I^- would be potential-determining ions. If such ions are responsible for the stabilization of a colloidal dispersion, they are referred to as peptizing ions. *See also* Indifferent Electrolyte.

Potential Energy Diagram   *See* Gibbs Energy of Interaction.

Potential Energy Barrier   *See* Activation Energy.

Potential Energy of Attraction   *See* Gibbs Energy of Attraction.

Potential Energy of Interaction   *See* Gibbs Energy of Interaction.

Potential Energy of Repulsion   *See* Gibbs Energy of Repulsion.

Pour Point   The lowest temperature at which an emulsion, oil, surfactant solution, or other material will flow under a standardized set of test conditions.

Power-Law Fluid   A fluid or dispersion whose rheological behavior is reasonably well described by the power-law equation. Here shear stress is set proportional to the shear rate raised to an exponent n, where n is the power-law index. The fluid is pseudoplastic for $n < 1$, Newtonian for $n = 1$, and dilatant for $n > 1$.

Primary Electroviscous Effect   *See* Electroviscous Effect.

Primary Minimum   In a plot of Gibbs energy of interaction versus separation distance, two minima may occur. The minimum occurring at the shortest distance of separation is referred to as the primary minimum, and that occurring at larger separation distance is termed the secondary minimum.

Primary Oil Recovery   The first phase of crude oil production, in which oil flows naturally to the wellbore. *See also* Secondary Oil Recovery, Enhanced Oil Recovery.

Probe Molecule   (Surfactant) Any species that is soluble in micelles and can be readily detected and measured. Example: pyrene solubilized in micelles can be a reporter probe for its environment through fluorescence spectroscopy.

Profilometry   *See* Scanning Tunneling Microscopy.

Promoter   In catalysis, any substance that interacts with a catalyst and causes an improvement in catalytic activity.

Protected Lyophobic Colloids   *See* Sensitization.

Protection   The process in which a material adsorbs onto droplet surfaces and thereby makes an emulsion less sensitive to aggregation and coalescence by any of a number of mechanisms. *See also* Sensitization.

Protective Colloid   A colloidal species that adsorbs onto and acts to "protect" the stability of another colloidal system. The term refers specifically to the protecting colloid and only indirectly to the protected colloid. Example: when a lyophilic colloid such as gelatin acts to protect another colloid in a dispersion by conferring steric stabilization. *See also* Protection.

PSD   (1) Particle size distribution.

(2) *See* Photon-Stimulated Desorption.

Pseudocolloid   A colloidal species or dispersion of colloidal species in which the colloidal unit has a an adsorbed component by which the colloid is detected or measured, as opposed to a "pure" colloidal species having no adsorbed material. Thus pseudocolloids (also termed Fremdkolloide or foreign colloid) have been distinguished from real or true colloids (also termed Eigenkolloide or self-colloid) although the use of all of these terms has been discouraged; *see* reference 18, page 72.

Pseudoemulsion Film   A fluid film of an aqueous phase (water) between air and oil phases. These are usually designated O/W/A or A/W/O. *See also* Fluid Film.

Pseudophase Diagram   A phase diagram for a system in which there are more phases present than are allowed to vary in the diagram. A pseudophase diagram is thus only one of several that are needed to completely describe a system.

Pseudoplastic   A non-Newtonian fluid whose viscosity decreases as the applied shear rate increases, a process that is also termed shear thinning. Pseudoplastic behavior may occur in the absence of a yield stress and also after the yield stress in a system has been exceeded (i.e., once flow begins).

Pulp   In mineral processing, a slurry of crushed ore dispersed in water.

Pure Colloid   *See* Pseudocolloid.

Pyruvic Acid Acetal   One of the kinds of polysaccharide structure comprising agar. Together with sulfated galactan, the combination is sometimes referred to as charged agar, or agaropectin. *See also* Agar.

QELS   Quasi-elastic light scattering. *See* Photon Correlation Spectroscopy.

Quadrupole   The property of having the equivalent of two dipoles (electric or magnetic), whose dipole moments have the same magnitude but point in opposite directions and are separated from each other.

Quasielastic Light Scattering (QELS)   *See* Light Scattering, Photon Correlation Spectroscopy.

Radiocolloid   Any colloidal species containing a radionuclide. Somewhat of a misnomer in that colloidal properties derive from the nature of atoms and molecules rather than from nuclei.

Rag Layer Emulsion   *See* Interface Emulsion.

Rain   *See* Atmospheric Aerosols of Liquid Droplets, Table 1.

Random Copolymer   A copolymer in which the constituent monomer molecules are randomly arranged in the polymer backbone. *See also* Block Copolymer.

Rayleigh–Gans–Debye Scattering   A modified model of Rayleigh scattering. *See also* Light Scattering.

Rayleigh Instability   *See* Rayleigh–Taylor Instability.

Rayleigh Ratio   In light scattering, the ratio of intenstites of incident to scattered light at some specified distance.

Rayleigh Scattering   Light will be scattered (deflected) by local variations in refractive index caused by the presence of dispersed species and depending upon their size. The scattering of light by species whose size is much less than the wavelength of the incident light is referred to as Rayleigh scattering, and it is termed Mie scattering if the species' size is comparable to that of the incident light. An example of Rayleigh scattering is that due to molecules in the atmosphere that scatter blue light from the sun's white-light illumination and cause the sky to appear blue while the sun appears orange–yellow. *See also* Mie Scattering, Light Scattering.

Rayleigh–Taylor Instability The instability of an interface between two fluids of different densities caused by the acceleration of the less dense fluid toward the more dense fluid.

RCF *See* Relative Centrifugal Force.

Real Colloid *See* Pseudocolloid.

Receding Contact Angle The dynamic contact angle that is measured when one phase is receding, or reducing its area of contact, along an interface while in contact with a third, immiscible phase. It is essential to state which interfaces are used to define the contact angle. *See also* Contact Angle.

Reduced Adsorption The relative Gibbs surface concentration of a component with respect to the total Gibbs surface concentration of all components. *See* reference 4 for the defining equations.

Reduced Limiting Sedimentation Coefficient *See* Sedimentation.

Reduced Osmotic Pressure *See* Colloid Osmotic Pressure.

Reduced Sedimentation Coefficient *See* Sedimentation.

Reduced Viscosity For solutions or colloidal dispersions, the specific increase in viscosity divided by the solute or dispersed-phase concentration, respectively ($\eta_{Red} = \eta_{SP}/C$). Also termed the viscosity number. *See* Table 4.

Refracted Light Light that has changed direction by passing from one medium through another in which its wave velocity is different.

Reflected Light Light that strikes a surface and is redirected back.

Reflection High-Energy Electron Diffraction (RHEED) *See* High-Energy Electron Diffraction.

Regioselective Catalyst   A catalyst that increases the rate or yield of a reaction when the reaction occurs at specific sites on the catalyst. Also termed regiospecific catalyst.

Relative Adsorption   The relative Gibbs surface concentration of a component with respect to that of another specified component. *See* reference 4 for the defining equations.

Relative Centrifugal Force (RCF)   When a centrifuge is used to enhance sedimentation or creaming, the centrifugal force is equal to mass times the square of the angular velocity times the distance of the dispersed species from the axis of rotation. The square of the angular velocity times the distance of the dispersed species from the axis of rotation, when divided by the gravitational constant, g, yields the relative centrifugal force or RCF. RCF is not strictly a force but rather the proportionality constant. It is substituted for g in Stokes' law to yield an expression for centrifuges and is used to compare the relative sedimentation forces achievable in different centrifuges. Because RCF is expressed in multiples of g, it is also termed g-force or simply gs.

Relative Micellar Mass   *See* Relative Molecular Mass of Micelles.

Relative Micellar Weight   *See* Relative Molecular Mass of Micelles.

Relative Molecular Mass   The mass of 1 mole of species (actually it is the mass of 1 mole of species divided by the mass of 1/12 mole of ^{12}C). Relative molecular mass has replaced the older term molecular weight. *See also* Number-Average Quantities, Mass-Average Quantities.

Relative Molecular Mass of Micelles   The mass of 1 mole of micelles (actually it is the mass of 1 mole of micelles divided by the mass of 1/12 mole of ^{12}C). Other terms that mean the same thing are relative micellar mass and relative micellar weight.

Relative Permittivity   *See* Permittivity, Table 7.

Relative Viscosity   In solutions and colloidal dispersions, the viscosity of the solution or dispersion divided by the viscosity of the solvent or continuous phase, respectively ($\eta_{Rel} = \eta/\eta_0$). Also termed the viscosity ratio. *See* Table 4.

Relative Viscosity Increment   *See* Specific Increase in Viscosity.

Relaxation Effect   *See* Electrophoretic Relaxation Effect.

Relaxation Time   The time required for the value of a changing property to be reduced to $1/e$ of its initial value.

Repeptization   Peptization, usually by dilution, of a once-stable dispersion that was aggregated (coagulated or flocculated) by the addition of electrolyte.

Replica   A metal film duplicate of a sample used in scanning electron microscopy. For example, an emulsion sample may be fast-frozen in a cryogen, fractured to reveal interior structure, then coated with a metal film to preserve the structure. See Freeze-Fracture Method.

Reporter Probe   *See* Probe Molecule.

Repulsive Potential Energy   *See* Gibbs Energy of Repulsion.

Retardation Effect   *See* Electrophoretic Retardation Effect.

Retarded van der Waals Constant   *See* van der Waals Constant.

Retentate   *See* Dialysis, Electrodialysis, Ultrafiltration.

Reverse Emulsion   A petroleum industry term used to denote an oil-in-water emulsion (most wellhead emulsions are W/O). Reverse emulsion is the opposite from the meaning of the term invert emulsion. *See also* Invert Emulsion.

Reverse Micelles   Synonym for the dispersed phase in a water-in-oil type microemulsion. Here the surfactant heads, or polar groups, associate closely to minimize interaction with the oil phase. This close association can happen when they orient themselves inside water droplets, and it also allows the surfactant tails, or hydrocarbon groups, to stabilize the water droplets by orienting toward or into the oil.

Reverse Osmosis   *See* Ultrafiltration.

RHEED   Reflection high-energy electron diffraction. *See* High-Energy Electron Diffraction.

Rheology   Strictly, the science of deformation and flow of matter. Rheological descriptions usually refer to the property of viscosity and departures from Newton's law of viscosity. *See also* Rheometer.

Rheomalaxis   A special case of time-dependent rheological behavior in which shear-rate changes cause irreversible changes in viscosity. The change can be negative, as when structural linkages are broken, or positive, as when structural elements become entangled (like work-hardening).

Rheometer   Any instrument designed for the measurement of non-Newtonian and Newtonian viscosities. The principal class of rheometer consists of the rotational instruments in which shear stresses are measured while a test fluid is sheared between rotating cylinders, plates, or cones. Examples of rotational rheometers: concentric cylinder, cone–cone, cone–plate, double-cone–plate, plate–plate, and disc. *See* reference 14.

Rheopexy   Dilatant flow that is time-dependent. At constant applied shear rate, viscosity increases, and in a flow curve, hysteresis occurs (but opposite to the thixotropic case).

Richardson Equation   (1) An empirical equation for emulsion viscosity. *See* Table 5.

(2) An expression giving the critical velocity needed for a liquid jet to break up into droplets and thus form an emulsion. *See* Ohnesorge Equation.

Rigid Film   *See* Fluid Film.

Ring–Roller Mill   A machine for the comminution, or size reduction, of minerals. Such machines crush the input material between a stationary ring and vertical rollers revolving inside the ring. Particle sizes of as low as about 30 μm can be produced.

Roll Crusher   A machine for the comminution, or size reduction, of mineral lumps or stones. Such machines crush the input material between a plate and revolving roller or between more than one roller.

Roll Mill   A device for imparting shear to a dispersion for the purpose of reducing the particle size of the dispersed material. Somewhat similar to a roll crusher. Examples of dispersions processed over a roll mill are pigment grinds (pigments dispersed in a fluid such as castor oil) and soap formulations (where the dispersed material includes fragrance oil droplets and pigments).

Ross Foam   Foam produced from a binary or ternary solution under conditions at which its temperature and composition approach (but do not reach) the point of phase separation into separate immiscible liquid phases.

Rotational Rheometer   *See* Rheometer.

Roughness Factor   The factor by which the surface area of a nonporous solid is greater than that calculated from the macroscopic dimensions of the surface.

Rupture   *See* Fluid Film.

Salinity Requirement   *See* Optimum Salinity.

Salt Curve   A graphical representation of the viscosity of a system versus salt concentration. This curve can be an important characteristic of formulated systems in which viscosity control is necessary, such as in shampoo formulas.

Salting In   Solutions: When the addition of electrolyte to a solution causes an increase in the solubility of a solute. *See also* Salting Out.

Surfactants: When the addition of electrolyte to a solution of nonionic surfactant causes the critical micelle concentration to increase. Also, addition of electrolyte to an ionic surfactant solution in a multiphase system can drive surfactant from the oil phase into the aqueous phase. *See also* Salting Out.

Salting Out   Solutions: When the addition of electrolyte to a solution causes a decrease in the solubility of a specified solute. *See also* Salting In.

Surfactants: When the addition of electrolyte to a solution of nonionic surfactant causes the critical micelle concentration to decrease. Also, addition of electrolyte to an ionic surfactant solution in a multiphase system can drive surfactant from the aqueous phase into the oil phase. *See also* Salting In.

Emulsions: The process of demulsification by the addition of electrolyte.

Sand   A term used to distinguish particles having different sizes in the range between about 50–63 µm and about 2000 µm, and with several subcategories, all depending on the operational scale adopted. *See* Table 3.

Scanning Electron Microscopy (SEM)   *See* Electron Microscopy.

Scanning High-Energy Electron Diffraction (SHEED)   *See* High-Energy Electron Diffraction.

Scanning Transmission Electron Microscopy (STEM)   *See* Electron Microscopy.

Scanning Tunneling Microscopy (STM)   A surface technique in which a metal probe, sharpened to a few tenths of a nanometer at the tip, is brought very close to a surface so that a current is generated resulting from overlapping wave functions (electron tunneling). The current varies with separation distance, and thus surface morphology may be determined as the tip is caused to move over the surface; the technique is referred to as profilometry. In an atomic force microscope (AFM), force rather than current is measured. *See also* Table 9.

Scattering Angle   *See* Light Scattering.

Scattering Plane   *See* Light Scattering.

Scavenging Flotation   A flotation separation process, in which particles or droplets become attached to gas bubbles that are injected (sparged) into the flotation medium. Also termed induced gas flotation. Example: the froth flotation of bitumen. *See also* Froth Flotation.

Schiller Layers   The layers of particles that may be formed during sedimentation such that the distances between layers are on the order of the wavelength of light, leading to iridescent, or Schiller layers.

Schlieren Optics   An optical arrangement designed to allow detection of density gradients occurring in fluid flow. Typically a narrow beam of light is collimated by one lens and focused on a knife-edge by a second lens. A density gradient in a fluid, between the lenses, causes a diffraction pattern to appear beyond the knife-edge.

Schulze–Hardy Rule   An empirical rule summarizing the general tendency of the critical coagulation concentration (CCC) of a suspension, an emulsion, or other dispersion to vary inversely with the sixth power of the counterion charge number of added electrolyte. *See also* Critical Coagulation Concentration.

Seaweed Colloids   A class of hydrophilic colloids (hydrocolloids) derived from various seaweeds. This class includes agar, algin, furcellaran, and carrageenan.

Secondary Electroviscous Effect   *See* Electroviscous Effect.

Secondary Ion Mass Spectroscopy (SIMS)   A technique for studying surface composition in which a beam of very high-energy ions strikes a surface and causes ions from the surface to be ejected. The masses of the ejected ions are determined in a mass spectrometer; hence surface compositions can be determined.

Secondary Minimum   *See* Primary Minimum.

Secondary Oil Recovery   The second phase of crude oil production, in which water or an immiscible gas are injected to restore production from a depleted reservoir. *See also* Primary Oil Recovery, Enhanced Oil Recovery.

Sediment   The process of sedimentation in a dilute dispersion generally produces a discernable, more concentrated dispersion that is termed the sediment and has a volume termed the sediment volume.

Sedimentation   The settling of suspended particles or droplets due to gravity or an applied centrifugal field. The rate of this settling is the sedimentation rate (or velocity). The sedimentation rate divided by acceleration is termed the sedimentation coefficient. The sedimentation coefficient extrapolated to zero concentration of sedimenting species is termed the limiting sedimentation coefficient. The sedimentation coefficient reduced to standard temperature and solvent is termed the reduced sedimentation coefficient. If extrapolated to zero concentration of sedimenting species it is termed the reduced limiting sedimentation coefficient. Negative sedimentation is also called flotation. Flotation in which droplets rise upwards is also called creaming. Flotation in which particulate matter becomes attached to gas bubbles is also referred to as froth flotation. *See also* Creaming, Froth Flotation, Subsidence.

Sedimentation Coefficient   *See* Sedimentation.

Sedimentation Equilibrium   The state of a colloidal system in which sedimentation and diffusion are in equilibrium.

Sedimentation Field Strength   *See* Sedimentation Potential.

Sedimentation Potential   The potential difference at zero current caused by the sedimentation of dispersed species. This mechanism of potential difference generation is known as the Dorn effect; accordingly, the sedimentation potential is sometimes referred to as the Dorn potential. The sedimentation may occur under gravitational or centrifugal fields. The potential difference per unit length in a sedimentation potential cell is the sedimentation field strength.

Sediment Volume   *See* Sediment.

Self-Colloid   *See* Pseudocolloid.

Self-Diffusion Coefficient   *See* Diffusion Coefficient.

SEM   Scanning electron microscopy. *See* Electron Microscopy.

Semi-Colloid   *See* Lyophilic Colloid.

Semipermeable Membrane   A membrane that is permeable to some species and not to others according to species size and electric charge. *See also* Colloid Osmotic Pressure.

Sensing-Zone Technique   A general term used to refer to any of the particle- or droplet-sizing techniques that rely on conductivity or capacitance changes in sample introduced between charged electrodes. An example is the Coulter counter.

Sensitization   The process in which small amounts of added hydrophilic colloidal material make a hydrophobic colloid more sensitive to coagulation by electrolyte. Example: the addition of polyelectrolyte to an oil-in-water emulsion to promote demulsification by salting out. Higher additions of the same material usually make the emulsion less sensitive to coagulation, and this is termed protective action or protection. The protected, colloidally stable dispersions that result in the latter case are termed protected lyophobic colloids.

Separator   In the petroleum industry, a vessel designed to separate the oil phase in a petroleum fluid from some or all of the other three constituent phases: gas, solids, and water. Free-water knockouts fall under this category, but so do separators capable of breaking and removing water and solids from emulsions. The latter range from gravity to impingement (coalescence) to centrifugal separators.

Septum   In general, any dividing wall between two cavities. Example: the thin liquid films (lamellae) between bubbles in a foam.

Sessile Bubble Method   *See* Sessile Drop Method.

Sessile Drop Method   A method for determining surface or interfacial tension based on measuring the shape of a droplet at rest on the surface of a solid substrate (in liquid–liquid systems the droplet may alternatively rest upside down, that is, underneath a solid substrate). This technique may also be used to determine the contact angle and contact diameter of the droplet against the solid.

Settling Radius   *See* Equivalent Spherical Diameter.

SEXAFS   Surface extended X-ray absorption fine structure spectroscopy. *See* Extended X-Ray Absorption Fine Structure Spectroscopy.

Shear   The rate of deformation of a fluid when subjected to a mechanical shearing stress. In simple fluid shear, successive layers of fluid move relative to each other such that the displacement of any one layer is proportional to its distance from a reference layer. The relative displacement of any two layers divided by their distance of separation from each other is termed the shear or the shear strain. The rate of change with time of the shear is termed the shear rate or the strain rate.

Shear Adhesion   *See* Peel Test.

Shear Plane   Any species undergoing electrokinetic motion moves with a certain immobile part of the electric double layer that is commonly assumed to be distinguished from the mobile part by a sharp plane, the shear plane. The shear plane is also termed the surface of shear. The zeta potential is the potential at the shear plane. *See also* Zeta Potential.

Shear Rate   *See* Shear.

Shear Stress   A certain applied force per unit area is needed to produce deformation in a fluid. For a plane area around some point in the fluid and in the limit of decreasing area the component of deforming force per unit area that acts parallel to the plane is the shear stress.

Shear Thickening   When the viscosity of a non-Newtonian fluid increases as the applied shear rate increases. *See also* Dilatant.

Shear Thinning   When the viscosity of a non-Newtonian fluid decreases as the applied shear rate increases. *See also* Pseudoplastic.

SHEED   Scanning high-energy electron diffraction. *See* High-Energy Electron Diffraction.

Sibree Equation   An empirical equation for the viscosity of emulsions. *See* Table 5.

Sieve   *See* Particle Size Classification.

Silicone Oil   Any of a variety of silicon-containing polymer solutions. An example is a linear poly(dimethylsiloxane): $HO[(CH_3)_2SiO]_nH$.

Silt   A term used to distinguish particles having sizes of greater than about 2–4 µm and less than about 50–63 µm, depending on the operational scale adopted. *See* Table 3.

Silver Film   *See* Black Film.

SIMS   *See* Secondary Ion Mass Spectroscopy.

Singer Equation   An equation for the surface pressure exerted by a polymer monolayer in terms of the area per moleucle and other properties of the polymer.

Sink-Float Method   *See* Hydrophobic Index.

Sintering   The coalescence or merging of two or more solid particles into a single particle.

Slip-Casting   In ceramics, the process in which a slurry of dispersed particles is poured into a mold, the liquid removed, and the particles sintered to form the final product.

Slurry Quality   In suspensions, the concentration of solid particles. In foams containing solids, the volume fraction of gas plus solid in the foam. *See also* Foam Quality.

Smectic State   *See* Mesomorphic Phase.

Smog   Aerosol air pollution, including dispersed liquids and solids having diameters less than about 2 μm.

Smoke   A special kind of aerosol that results from a thermal process such as combustion or thermal decomposition; the aerosol may be of solid particles or of liquid droplets.

Smoluchowski Equation   Electrophoresis: A relation expressing the proportionality between electrophoretic mobility and zeta potential for the limiting case of a species that can be considered to be large and have a thin electric double layer. Also termed Helmholtz–Smoluchowski Equation. *See also* Hückel Equation, Henry Equation, Electrophoresis.

   Diffusion: A relation expressing the proportionality between the rate of diffusional encounters between species and their diffusion coefficient, in which the constant of proportionality includes the radius and number concentration of the species. There are also derived Smoluchowski equations for specific processes, such as for rate of aggregation of particles.

Snap-Off   A mechanism for foam lamella generation in porous media. When gas enters and passes through a constriction in a pore, a capillary pressure gradient is created, and it causes liquid to flow toward the region of the constriction, where it accumulates and may cause the gas to pinch off or snap off and thereby create a new gas bubble separated from the original gas by a liquid lamella. *See also* Lamella Division, Lamella Leave-Behind.

Soap   A surface-active fatty acid salt containing at least eight carbon atoms. The term is no longer restricted to fatty acid salts originating from natural fats and oils. *See also* Surfactant.

Soap Curd   A mixture of soap crystals in a saturated solution in which the soap crystals produce a gel-like consistency. The soap crystals in this case are referred to as curd-fibers. Soap curd is not a mesomorphic (liquid-crystal) phase.

Soap Film   A thin film of water in air that is stabilized by surfactant. The term is used even though the film is not a film of soap and even where the surfactant is not a soap. *See also* Fluid Film.

Soft X-Ray Appearance Potential Spectroscopy (SXAPS)
See Appearance Potential Spectroscopy.

Soft X-Ray Emission Spectroscopy (SXES)   *See* X-Ray Emission Spectroscopy.

Soil   Naturally occurring unconsolidated material, whether mineral or organic, that is on the earth's surface and is capable of supporting plant growth.

Sol   A colloidal dispersion. In some usage the term sol is used to distinguish dispersions in which the dispersed-phase species are of very small size so that the dispersion appears transparent.

Solid Aerosol   *See* Aerosol of Solid Particles.

Solid Emulsion   A colloidal dispersion of a liquid in a solid. Examples: opal, pearl.

Solid Foam   A colloidal dispersion of a gas in a solid. Example: polystyrene foam.

Solidification Front Method   A method for determining solid surface or interfacial tension. When a small particle, initially embedded in a liquid phase, is slowly approached by the solid–liquid interface of the slowly solidifying (freezing) liquid, it will remain embedded or become rejected by the front depending upon the free energy of adhesion, which in turn depends upon the solid–liquid interfacial tension. Also termed the freezing front method. *See also* the review by Spelt (reference 46).

Solid Suspension   A colloidal dispersion of a solid in another solid. Example: ruby-stained glass, a dispersion of gold particles in glass.

Solubility Parameter   A means of estimating solubility. It is the square root of the ratio of energy of vaporization to molar volume. This ratio is also known as the cohesive energy density. Pairs of substances having very similar solubility parameters tend to be mutually soluble.

Solubilizate   The solute whose solubility is increased in the process of solubilization.

Solubilization   The process by which the solubility of a solute is increased by the presence of another solute. Micellar solubilization refers to the incorporation of a solute (solubilizate) into or on micelles of another solute to thereby increase the solubility of the first solute.

Solvent-Motivated Sorption   Sorption that occurs as a result of the hydrophobicity of a compound. Accumulation of the compound at the interface or in the other phase is not due to its affinity for that phase so much as to its disaffinity for the initial phase. Such sorption occurs for organic contaminants in the environment. This kind of sorption can often be related to the octanol–water partitioning coefficient. *See also* Hydrophobic Organic Contaminant, Sorbent-Motivated Sorption.

Solvophoresis   A variant of diffusiophoresis. If a particle is immersed into a mixed solvent in which a concentration gradient exists, the particle will tend to move in the direction of increasing concentration of the solvent component that is preferentially adsorbed onto its surface (reference 36).

Sorbate   A substance that becomes sorbed into an interface or another material or both. *See also* Sorption.

Sorbent   The substrate into which or onto which a substance is sorbed or both. *See also* Sorption.

Sorbent-Motivated Sorption   Sorption that occurs as a result of the affinity of the surface for a particular compound. Example: ion exchange. *See also* Solvent-Motivated Sorption.

Sorption   A term used in a general sense to refer to either or both of the processes of adsorption and absorption.

Sorptive   Material that is present in one or both of the bulk phases bounding an interface and capable of becoming sorbed.

Specific Increase in Viscosity   The relative viscosity minus unity. Also referred to as specific viscosity or relative viscosity increment. *See* Table 4.

Specific Surface Area   *See* Surface Area.

Specific Viscosity   *See* Specific Increase in Viscosity.

Spherical Agglomeration   The process of separating particles from their suspension by selective wetting and agglomeration with a second, immiscible liquid. The second liquid preferentially wets the particles and causes particle adhesion by capillary liquid bridges (reference 37). The process of agglomeration thus includes both aggregation and coalescence. *See also* Agglomeration.

Spinning-Drop Method   A method for determining surface or, more commonly, interfacial tension based on measuring the shape of a droplet (or bubble) suspended in the center of a horizontal, cylindrical tube filled with a liquid while the tube is spinning about its long axis. This method is particularly suited to the determination of very low interfacial tensions.

Spray   A dispersion of a liquid in a gas (aerosol of liquid droplets) in which the droplets have diameters greater than 10 μm. *See also* Mist.

Spread Monolayer   *See* Spread Layer.

Spread Layer   The interfacial layer formed by an adsorbate when it becomes essentially completely adsorbed out of the bulk phase(s). If the layer is known to be one molecule thick, then the term spread monolayer is used.

Spreading   The tendency of a liquid to flow and form a film coating an interface, usually a solid or immiscible liquid surface, in an attempt to minimize interfacial free energy. Such a liquid forms a zero contact angle as measured through itself.

Spreading Coefficient   A measure of the tendency for a liquid to spread over a surface (usually of another liquid). It is −1 times the Gibbs free energy change for this process (the work of adhesion between the two phases minus the work of cohesion of the spreading liquid) so that spreading is thermodynamically favored if the spreading coefficient is greater than zero. In a gas–liquid system containing a potentially spreading liquid A, a substrate liquid L, and gas, the spreading coefficient is given by $S = \gamma°_L - \gamma_{L/A} - \gamma°_A$, where $\gamma°_L$ and $\gamma°_A$ are surface tensions and $\gamma_{L/A}$ is interfacial tension. Early usage of the concept involved terms such as the spreading parameter or wetting power (reference 38). When equilibria at the interfaces are not achieved instantaneously, reference is frequently made to the initial spreading coefficient and final (equilibrium) spreading coefficient. If the initial spreading coefficient is positive and the final spreading coefficient negative, the system exhibits autophobicity. *See also* Entering Coefficient.

Spreading Parameter   *See* Spreading Coefficient.

Spreading Pressure   *See* Surface Pressure.

Spreading Tension   A synonym for spreading coefficient.

Spreading Wetting   The process of wetting in which a liquid, already in contact with a solid (or second, immiscible liquid) surface, spreads over the solid surface, thereby increasing the interfacial area of contact between them. The spreading is thermodynamically favored when the spreading coefficient is positive. *See also* Adhesional Wetting, Immersional Wetting, Spreading Coefficient, Wetting.

Stability   *See* Colloid Stability, Thermodynamic Stability.

Stalagmometric Method   *See* Drop-Weight Method.

Standard Sedimentation Coefficient   A synonym for reduced sedimentation coefficient. *See* Sedimentation.

Static Coefficient of Friction   *See* Friction.

Static Interfacial Tension   *See* Static Surface Tension.

Static Mixer   A device for mixing components in a solution or dispersion without moving mechanical elements. Stationary flow guiding elements are built into a device, frequently a section of pipe, and induce mixing and dispersion by repeatedly dividing and recombining partial streams of the flowing material. Also termed motionless mixers.

Static Surface Tension   A synonym for the equilibrium surface tension or interfacial tension. *See* Surface Tension.

Staudinger–Mark–Houwink Equation   An empirical equation relating intrinsic viscosity to molecular mass for polymer solutions. *See* Table 5. This equation may used to determine viscosity-average molecular mass. Sometimes referred to as the Mark–Houwink Equation.

STEM   Scanning transmission electron microscopy. *See* Electron Microscopy.

Steric Stabilization   The stabilization of dispersed species induced by the interaction (steric stabilization) of adsorbed polymer chains. Example: adsorbed proteins stabilize the emulsified oil (fat) droplets in milk by steric stabilization. Also termed depletion stabilization. *See also* Protection.

Stern Layer   The layer of ions in an electric double layer that, hydrated or not, lie adjacent to the surface (adsorbed ions). The rest of the electric double layer is often distinguished as the diffuse part, where assumptions such as that treating the ions as point charges can more reasonably be made. *See also* Electric Double Layer, Helmholtz Double Layer.

Stern Layer Potential   *See* Inner Electrical Potential.

Stern Potential   *See* Inner Electrical Potential.

STM   *See* Scanning Tunneling Microscopy.

Stokes' Law   A relation giving the terminal settling velocity of a sphere as $2r^2 \Delta \rho g/(9\eta)$, where r is the sphere radius, $\Delta \rho$ is the density difference between the phases, g is the gravitational constant, and η is the external-phase viscosity.

Stokes Diameter   *See* Equivalent Spherical Diameter.

Strain, Strain Rate   *See* Shear Rate.

Stratified Film   A fluid film in which several thicknesses can exist simultaneously and can persist for a significant amount of time. *See also* Fluid Film.

Streaming Current   In electrokinetic motion, the current due to relative displacement of the part of the electric double layer beyond the shear plane. The electric current flowing in a streaming potential cell when the electrodes are short-circuited.

Streaming Potential   The potential difference at zero current created when liquid is made to flow through a porous medium.

Streamline Flow   *See* Laminar Flow.

Stress   *See* Shear Stress.

Stress Relaxation   *See* Viscoelastometer.

Submicron   An older particle size range distinction no longer in use. *See also* Micrometer, Micron.

Subsidence   The process of sedimentation in which the settling of suspended particles results in a dense compaction, or coagulation, of particles in which liquid is squeezed out. Geologically, significant compaction of clay layers due to lowering of the water table (dewatering).

Substrate   A material that provides a surface or interface at which adsorption or other phenomena take place.

Suction Pressure   *See* Capillary Pressure.

Sugden's Parachor   *See* Parachor.

Sulfated Galactan   One of the kinds of polysaccharide structure that constitutes agar. Together with pyruvic acid acetal, the combination is sometimes referred to as charged agar, or agaropectin. *See also* Agar.

Supercentrifuge   *See* Centrifuge.

Superficial Density   An older term now replaced by the Gibbs surface concentration, or simply, the surface excess.

154   THE LANGUAGE OF COLLOID AND INTERFACE SCIENCE

Surface   *See* Interface.

Surface-Active Agent   *See* Surfactant.

Surface Area   The area of a surface or interface, especially that between a dispersed and a continuous phase. The specific surface area is the total surface area divided by the mass of the appropriate phase.

Surface Charge   The fixed charge that is attached to, or part of, a colloidal species' surface and forms one layer in an electric double layer. There is thus a surface-charge density associated with the surface. *See also* Electric Double Layer.

Surface-Charge Density   *See* Surface Charge.

Surface Concentration   *See* Gibbs Surface.

Surface Conductivity   The excess conductivity, relative to the bulk solution, in a surface or interfacial layer per unit length. Also termed the surface excess conductivity.

Surface Coverage   The ratio of the amount of adsorbed material to the monolayer capacity. The definition is the same for either of monolayer and multilayer adsorption.

Surface Dilational Modulus   *See* Film Elasticity.

Surface Dilational Viscosity   *See* Surface Viscosity.

Surface Elasticity   *See* Film Elasticity.

Surface Excess (Concentration)   *See* Gibbs Surface.

Surface Excess Conductivity   *See* Surface Conductivity.

Surface Excess Isotherm   A function relating, at constant temperature and pressure, the relative adsorption or reduced adsorption, or similar surface excess quantity to the concentration of component in the equilibrium bulk phase.

Surface-Extended X-Ray Absorption Fine Structure Spectroscopy (SEXAFS)   *See* Extended X-Ray Absorption Fine Structure Spectroscopy.

Surface Fluidity   The inverse of the surface shear viscosity.

Surface Layer   *See* Interfacial Layer.

Surface Layer of Adsorbent   *See* Adsorption Space.

Surface of Shear   *See* Shear Plane.

Surface of Tension   An imaginary boundary, having no thickness, at which surface or interfacial tension acts.

Surface Potential   The potential at the interface bounding two phases, that is, the difference in outer (Volta) potentials between the two phases. *See also* Inner Potential, Outer Potential, Jump Potential.

Surface Potential Jump   *See* Jump Potential.

Surface Pressure   Actually an analog of pressure; the force per unit length exerted on a real or imaginary barrier separating an area of liquid or solid that is covered by a spreading substance from a clean area on the same liquid or solid. Also referred to as spreading pressure.

Surface Rheology   *See* Surface Viscosity.

Surface Rheometer   *See* Surface Viscometer.

Surface Shear Viscosity   *See* Surface Viscosity.

Surface Tension   The contracting force per unit length around the perimeter of a surface is usually referred to as surface tension if the surface separates gas from liquid or solid phases, and interfacial tension if the surface separates two nongaseous phases. Although not strictly defined the same way, surface tension can be expressed in units of energy per unit surface area. For practical purposes surface tension is frequently taken to reflect the change in surface free energy per unit increase in surface area. *See also* Surface Work.

Surface Viscosity   The two-dimensional analog of viscosity acting along the interface between two immiscible fluids. Also called interfacial viscosity. In fact, there are two kinds of surface viscosity: surface shear viscosity and surface dilational viscosity. Surface shear viscosity is the component that is analogous to three-dimensional shear viscosity: the rate of yielding of a layer of fluid due to an applied stress. Surface dilational viscosity relates to the rate of area expansion and is expressed as the local gradient in surface tension per change in relative area per unit time. Any shear rate dependence (non-Newtonian behavior) falls under the subject of surface rheology. Although usually termed surface viscosity or rheology, especially when one fluid is a gas, the more general terminology is surface or interfacial rheology. *See also* Viscosity.

Surface Viscometer   An instrument for determining surface rheological properties. One such type of instrument operates by rotating a ring or disk to apply shear in the plane of the interface, while maintaining the area constant. Alternatively, measurements are made by expanding or contracting the interface. *See also* Canal Viscometer, Torsional Viscometer.

Surface Work   The work required to increase the area of the surface of tension. Under reversible, isothermal conditions the surface work (per unit surface area) equals the equilibrium, or static, surface tension.

Surfactant   Any substance that lowers the surface or interfacial tension of the medium in which it is dissolved. The substance does not have to be completely soluble and may lower surface or interfacial tension by spreading over the interface. Soaps (fatty acid salts containing at least eight carbon atoms) are surfactants. Detergents are surfactants, or surfactant mixtures, whose solutions have cleaning properties. Also referred to as surface-active agents or tensides. In some usage surfactants are defined as molecules capable of associating to form micelles.

Surfactant Effectiveness   The surface excess concentration of surfactant corresponding to saturation of the surface or interface. Example: one indicator of effectiveness is the maximum reduction in surface or interfacial tension achievable by a surfactant. This term has a different meaning from surfactant efficiency. *See* references 15 and 16.

Surfactant Efficiency   The equilibrium solution surfactant concentration needed to achieve a specified level of adsorption at an interface. Example: one such measure of efficiency is the surfactant concentration needed to reduce the surface or interfacial tension by 20 mN/m from the value of the pure solvent(s). This term has a different meaning from surfactant effectiveness. *See* references 15 and 16.

Surfactant Tail   The lyophobic portion of a surfactant molecule. It is commonly a hydrocarbon chain containing eight or more carbon atoms. *See also* Head Group.

Suspended Sediment   The insoluble particulate matter in natural water bodies such as rivers, lakes, and oceans.

Suspending Power   The ability of a detergent or detergent component to keep foreign material away from the solid material from which it has been removed in order to prevent redeposition. *See also* Detergency, Detergent.

Suspension   A system of solid particles dispersed in a liquid. Suspensions were previously referred to as suspensoids, meaning suspension colloids. Aside from the obvious definition of a colloidal suspension, a number of operational definitions are common in industry, such as any dispersed matter that can be removed by a 0.45 μm nominal pore size filter.

Suspension Effect   A finite potential, the Donnan potential, may exist between a suspension and its equilibrium solution. Also referred to as the Pallmann or Wiegner effect.

Svedberg   A unit of the sedimentation coefficient equal to 10^{-13} s.

Swamping Electrolyte   An excess of indifferent electrolyte that severely compresses electric double layers and minimizes the influence of electric charges borne by large molecules or dispersed colloidal species.

Swelling   The increase in volume associated with the uptake of liquid or gas by a solid or a gel.

Swelling Pressure   The pressure difference between a swelling material and the bulk of fluid being imbibed that is needed to prevent additional swelling. *See also* Swelling.

Swollen Micelle   *See* Micelle.

SXAPS   Soft X-ray appearance potential spectroscopy. *See* Appearance Potential Spectroscopy.

SXES   Soft X-ray emission spectroscopy. *See* X-Ray Emission Spectroscopy.

Syndet   A synthetic detergent other than a soap.

Syndiotactic Polymer   *See* Atactic Polymer.

Syneresis   The spontaneous shrinking of a colloidal dispersion due to the release and exudation of some liquid. Frequently occurs in gels and foams but also occurs in flocculated suspensions. Mechanical syneresis refers to enhancing syneresis by the application of mechanical forces.

Synergistic Electrolyte   *See* Critical Coagulation Concentration.

Szyszkowski Equation   An equation for estimating the surface tensions of aqueous solutions of various concentrations. *See* Table 8.

Tactoid   (1) In the destabilization of lyophilic colloids when coacervation occurs, the dispersed phase may initially separate into small, anisotropic droplets having shapes such as cylinders, called tactoids. With concentrated colloids, droplets of dilute colloid may separate out within the concentrated colloid; these droplets are sometimes referred to as negative tactoids.

(2) In clay suspensions the thin sheetlike or platelike particles may aggregate to form stacks of particles in face-to-face orientation, which are termed tactoids.

TAN   *See* Total Acid Number.

Tate's Law   An equation giving the mass of a droplet that forms and falls from a capillary tube of radius r, as the product of the surface tension times $2\pi r$. This equation is not very accurate, and significant corrections must be applied. *See also* Drop-Weight Method.

Taylor Equation   An empirical equation for estimating the viscosity of an emulsion. *See* Table 5.

TDS   (1) Total dissolved solids.

(2) Thermal desorption spectroscopy. *See* Temperature-Programmed Reaction Spectroscopy.

TEM   Transmission electron microscopy. *See* Electron Microscopy.

Temperature-Programmed Reaction Spectroscopy (TPRS)
A surface technique in which thermally stimulated adsorbed species
are desorbed from a surface in order to gain information about
adsorbate–substrate bonding and about surface composition. This
technique is destructive in that the heating drives off the adsorbed
species. Depending on whether the temperature rise is conducted
quickly or slowly, two techniques are distinguished: flash desorption
spectroscopy (FDS) and thermal desorption spectroscopy (TDS),
also termed temperature-programmed desorption spectroscopy
(TPDS). *See also* Photon-Stimulated Desorption Spectroscopy.

Tenside   *See* Surfactant.

Tensiometer   A general term applied to any instrument that
may be used to measure surface and interfacial tension.

Tertiary Electroviscous Effect   *See* Electroviscous Effect.

Tertiary Oil Recovery   *See* Enhanced Oil Recovery.

Thermal Desorption Spectroscopy (TDS)   *See* Temperature-
Programmed Reaction Spectroscopy.

Thermionic Work Function   The work needed to move an electron from the highest occupied level in a metal to a position outside
the metal.

Thermodynamic Stability   In colloid science, the terms thermodynamically stable and metastable mean that a system is in a
state of equilibrium corresponding to a local minimum of free energy (reference 4). If several states of energy are accessible, the lowest
is referred to as the stable state and the others are referred to as
metastable states; unstable states are not at a local minimum. Most
colloidal systems are metastable or unstable with respect to the separate bulk phases. *See also* Colloid Stability, Kinetic Stability.

Thermotropic Liquid Crystals   *See* Mesomorphic Phase.

Thermotropic Mesomorphic Phase   *See* Mesomorphic Phase.

Theta Temperature   The temperature at which a polymer solution exhibits ideal solution behavior, polymer–polymer attraction
equals polymer–solvent attraction, and polymer dissolution is due
to entropy change alone. Also referred to as Flory point and Flory
temperature.

Thickness of the Electric Double Layer   *See* Electric-Double-Layer Thickness.

Thin Film   *See* Fluid Film.

Thin-Film Drainage   *See* Film Drainage, Fluid Film.

Thixotropic   Pseudoplastic flow that is time-dependent. At constant applied shear rate, viscosity gradually decreases, and in a flow curve hysteresis occurs. That is, after a given shear rate is applied and then reduced, it takes some time for the original dispersed species' alignments to be restored.

Thomas Equation   An empirical equation for estimating the viscosity of a dispersion. *See* Table 5.

Three-Phase Separator   *See* Separator.

Tight Emulsion   A petroleum industry term for a practically stable emulsion, in contrast to a less stable, or "loose" emulsion.

Tilting-Plate Method   A means of determining the contact angle between a solid plate and the liquid into which it is immersed. The plate is adjusted to various angles until a flat (horizontal) meniscus is obtained, in which case the plate angle measured through the liquid yields the desired contact angle.

Tiselius Apparatus   An apparatus for the determination of electrophoretic mobilities. *See* Moving Boundary Electrophoresis.

Topochemical Reaction   A chemical reaction that can not be expressed stoichiometrically. Such reactions may occur only at certain locations on a molecule or only for certain molecular orientations.

Torsional Viscometer   An instrument for the qualitative determination of surface or interfacial viscosity in which a circular flat or double-cone bob, placed in the plane of and just contacting the surface or interface, is caused to oscillate. With the knife-edge supported by a torsion wire, the damping of oscillations is observed. Alternatively, in a two-dimensional analog of the concentric cylinder rheometer, constant rotational shear is applied and torque is measured.

Tortuosity   In porosimetry evaluations, experimental data tend to be interpreted in terms of a model in which the porous medium is taken to comprise a bundle of cylindrical pores having radius r. If the Young–Laplace equation is then applied to the data, an effective value of r can be calculated, even though this model ignores the real distribution of irregular channels. The calculated r value is sometimes considered to represent the radius of an equivalent cylinder or, alternatively, is termed the tortuosity.

Total Acid Number (TAN)   The acid number expresses the amount of base (potassium hydroxide) that will react with a given amount of material in a standardized titration procedure. A large acid number indicates a high concentration of acids in the original material, usually including natural surfactant precursors. A commonly measured property of crude oils.

Total Potential Energy of Interaction   *See* Gibbs Energy of Interaction.

TPRS   *See* Temperature-Programmed Reaction Spectroscopy.

Tracer Diffusion Coefficient   The diffusion coefficient of an isotopically labeled species. Usually taken to be equal to the diffusion coefficient of the corresponding unlabeled species. *See also* Diffusion Coefficient.

Transitional Pore   An older term now replaced by mesopore. *See* Pore.

Transmission Electron Microscopy (TEM)   *See* Electron Microscopy.

Traube's Rule   A generalization for homologous series of organic compounds of type $R(CH_2)_nX$, that for each incremental CH_2 group the concentration of molecules required to produce a specified surface tension decreases by a factor of about 3. In adsorption Traube's rule is that a polar adsorbent will preferentially adsorb the most polar component from a nonpolar solution, and conversely, a nonpolar adsorbent will preferentially adsorb the least polar component from a polar solution.

Treater   A vessel used for the breaking of emulsions and the consequent removal of solids and water (BS&W). Emulsion breaking may be accomplished through some combination of thermal, electrical, chemical, or mechanical methods. A treater might be applied to break an emulsion and separate solids and water that could not be removed in a separator.

Tribology   The science of friction and lubrication.

True Colloid   *See* Pseudocolloid.

Turbidimetry   *See* Turbidity.

Turbidity   The property of dispersions that causes a reduction in the transparency of the continuous phase due to light scattering and absorption. Turbidity is a function of the size and concentration of the dispersed species. The turbidity coefficient is simply the extinction coefficient in the Beer–Lambert equation for absorbance when light scattering rather than absorbance proper is being studied (hence turbidimetry). *See also* Nephelometry.

Turbulent Flow   A condition of flow in which all components of a fluid passing a certain point do not follow the same path. Turbulent flow refers to flow that is not laminar, or streamline.

Tyndall Beam   *See* Tyndall Scattering.

Tyndall Scattering   A process that produces a colored beam of light scattered by uniform dispersion of particles whose size approaches the wavelength of the incident light. The scattered light is referred to as a Tyndall beam. The wavelength of scattered light varies with the angle of observation (Tyndall spectra), and this feature allows particle size to be calculated. *See also* Higher Order Tyndall Spectra.

Ubbelohde Viscometer   *See* Capillary Viscometer.

Ultracentrifuge   *See* Centrifuge.

Ultrafiltrate   *See* Ultrafiltration.

Ultrafiltration   A separation process somewhat like dialysis in which a colloidal dispersion is separated from a noncolloidal solution by a semipermeable membrane, that is, a membrane that is permeable to all species except the colloidal-sized ones. Here an applied pressure (rather than osmotic pressure) across the membrane drives the separation. As in dialysis the solution containing the colloidal species is referred to as the retentate or dialysis residue. However, the solution that is free of colloidal species is referred to as ultrafiltrate rather than dialysate, because the composition is usually different from that produced by dialysis. Also referred to as hyperfiltration or reverse osmosis. *See also* Dialysis.

Ultramicroscope   An optical microscope that uses dark-field illumination to make visible extremely small (submicrometer-sized) particles or droplets. Also termed Dark-Field Microscope. *See also* Dark-Field Illumination.

Ultrasonic Dispersion   The use of ultrasound waves to achieve or aid in the dispersion of particles or droplets.

Ultrasound Vibration Potential (UVP)   The electrokinetic potential of colloidal species detected by the electric field generated when the species are made to move by an imposed ultrasonic field. This method can be applied to W/O emulsions that do not transmit light. For low potentials, the UVP can be quantitatively related to the electrophoretic mobility.

Ultraviolet Photoelectron Spectroscopy (UPS)   *See* Photoelectron Spectroscopy.

Ultraviolet Photoemission Spectroscopy (UPS)   *See* Photoelectron Spectroscopy.

Unactivated Adsorption   Physisorption, that is, adsorption for which there is no activation energy barrier to be overcome, as opposed to activated adsorption, or chemisorption, for which an activation energy barrier must be overcome. See also Chemisorption, Physisorption.

Upper-Phase Microemulsion   A microemulsion with a high oil content that is stable while in contact with a bulk water phase and in laboratory tube or bottle tests tends to be situated at the top of the tube above the water phase. For chlorinated organic liquids, which are more dense than water, the oil will be the top phase rather than the bottom. *See also* Microemulsion, Winsor Type Emulsions.

Upper Plastic Limit   *See* Liquid Limit.

UPS   Ultraviolet photoelectron spectroscopy or ultraviolet photoemission spectroscopy. *See* Photoelectron Spectroscopy.

U.S. Bureau of Mines Wettability Test   *See* Wettability Index.

UVP   *See* Ultrasound Vibration Potential.

van der Waals Adsorption   An older term now replaced by physical adsorption, or physisorption. *See also* Chemisorption.

van der Waals Forces   *See* Dispersion Forces.

van der Waals–Hamaker Constant   *See* Hamaker Constant.

van't Hoff's Law   The relation specifying that the osmotic pressure of a solution equals the gas pressure the solute would exert if it were an ideal gas occupying the same volume as the solution.

Velocity Gradient   A parameter that indicates the intensity of mixing. It is a function of the power input, the reactor volume, and the fluid viscosity. Higher velocity gradients are used in coagulation where the goal is to disperse the coagulant to the particle surfaces. Lower velocity gradients are used in flocculation where the goal is particle collisions and aggregation, and higher gradients would break up flocs.

Versator   A device that is used for deaerating liquid systems, such as emulsions, and operates on the principle of centrifugally generating a thin film of the liquid with high shear and exposing the thin film to vacuum.

Very Coarse Sand   *See* Sand, Table 3.

Very Fine Sand   *See* Sand, Table 3.

Vesicle   A droplet that is stabilized by the presence at its surface of a lipid bimolecular film (bilayer) or series of concentric bilayers. Also termed liposome. *See also* Bimolecular Film.

Virial Equation   An equation in which there is a power series of terms of increasing order in the independent variable. Each term has associated with it a coefficient termed its virial coefficient. An example of a virial equation is the nonideal gas law when written as a power series. The second virial coefficient describes the first order of deviations from nonideality.

Viscoelastic   A liquid (or solid) with both viscous and elastic properties. A viscoelastic liquid will deform and flow under the influence of an applied shear stress, but when the stress is removed the liquid will slowly recover from some of the deformation.

Viscoelastometer   An instrument for studying viscoelastic fluids. Viscoelastometers may be used to apply a constant shear stress so that the resulting deformation can be determined (creep curve), or to apply a sudden deformation and determine the stress needed to maintain the deformation (stress relaxation).

Viscoelectric Constant   A reflection of the increase in viscosity of a liquid due to the presence of an electric field. It is given by the increase in viscosity divided by the viscosity in the absence of an electric field, and divided also by the square of the electric-field gradient.

Viscometer   Any instrument employed in the measurement of viscosity. In most cases the term is applied to instruments capable of measuring only Newtonian viscosity and not capable of non-Newtonian measurements. *See* Rheometer.

Viscosimeter   *See* Viscometer.

Viscosity   A measure of the resistance of a liquid to flow. It is properly the coefficient of viscosity, and expresses the proportionality between shear stress and shear rate in Newton's law of viscosity. For variations, *see* Table 4. Many equations have been used to predict the viscosities of colloidal dispersions; *see* Table 5.

Viscosity-Average Molecular Mass   Molecular mass determined on the basis of viscosity measurements coupled with an empirical equation such as the Staudinger–Mark–Houwink equation.

Viscosity Number   *See* Reduced Viscosity.

Viscosity Ratio   *See* Relative Viscosity.

Volta Potential   *See* Outer Potential.

Votator   A continuous-process device for rapidly changing the temperature of a liquid system. Liquid enters the device, is spread in a thin film over a heat-exchanging surface, is then removed from the surface by wall scrapers, and then exits the device through an outlet.

Wagner Equation   An equation for predicting the conductivity of dispersions. *See* Table 6.

Washburn Equation   An equation describing the extent of displacement of one fluid by another in a capillary tube or cylindrical pore in a porous medium. If h is the depth of penetration of invading fluid and dh/dt is the rate of penetration, then $dh/dt = \gamma r \cos \theta/(4\eta h)$, where γ is the interfacial tension, r is the capillary radius, θ is the contact angle, and η is the viscosity of the invading fluid. It is used in the evaluation of porosimetry data and may be used to provide information about contact angles, capillary radii, and pore radii, depending on the experiments conducted.

Water II   *See* Polywater.

Wet   *See* Wettability.

Wet Oil   An oil containing free water or emulsified water.

Wettability   A qualitative term referring to the water- or oil-preferring nature of surfaces, such as mineral surfaces. Example: the flow of emulsions in porous media is influenced by the wetting state of the walls of pores and throats through which the emulsion must travel. Wettability may be determined by direct measurement of contact angles, or inferred from measurements of fluid imbibition or relative permeabilities. Several conventions for describing wettability values exist. *See also* Amott Test, Contact Angle, Wettability Index, Wetting.

Wettability Index   A measure of wettability based on the U.S. Bureau of Mines (USBM) wettability test in which the wettability index (W) is determined as the logarithm of the ratio of areas under the capillary pressure curves for both increasing and decreasing saturation of the wetting phase. Complete oil-wetting occurs for $W = -\infty$ (in practice about –1.5), and complete water-wetting occurs for $W = \infty$ (in practice about 1.0). Another wettability index is derived from the Amott–Harvey test. *See also* reference 21, Wettability, Amott Test.

Wetting   A general term referring to one or more of the following specific kinds of wetting: adhesional wetting, spreading wetting, and immersional wetting. Frequently used to denote that the contact angle between a liquid and a solid is essentially zero and there is spontaneous spreading of the liquid over the solid. Nonwetting, on the other hand, is frequently used to denote the case where the contact angle is greater than 90° so that the liquid rolls up into droplets. *See also* Draves Wetting Test, Contact Angle, Wettability.

Wetting Coefficient   In the equilibrium of a system of solid, gas, and liquid, the wetting coefficient, k, is given as $k = (\gamma_{sg} - \gamma_{sl})/\gamma_{lg}$, where γ is the interfacial tension and the subscripts g, l, and s refer to gas, liquid, and solid phases, respectively, at the interfaces. Complete wetting occurs for $k \geq 1$ and nonwetting for $k \leq 1$.

Wetting Power   *See* Spreading Coefficient.

Wetting Tension   The work done on a system during the process of immersional wetting, expressed per unit area of the phase being immersionally wetted. *See also* Immersional Wetting.

Whey   A term in dairy processing referring to the dilute oil-in-water emulsion that separates from the coagulated portion, or curd, in cheese-making.

Whey-Off   A term in dairy processing referring to any unwanted process in which whey separates from a product.

Whipping Agent   *See* Foaming Agent.

Wiegner Effect   *See* Suspension Effect.

Wiener Equation   For predicting the relative permittivity of dispersions. *See* Table 7.

Wilhelmy Plate Method   A method for determining surface or interfacial tension based on measuring the force needed to pull an inert plate, held vertically, through an interface. Also termed the Wilhelmy slide method. *See also* du Nouy Ring Method.

Wine Tears   The layer of droplets that can appear at the top of the meniscus in a container of an alcohol–water solution (e.g., wine). The pumping action that draws liquid up through the meniscus is a result of the evaporation of alcohol from a thin region at the top of the layer, raising the surface tension and causing liquid to rise from the bulk into the layer (Marangoni flow). In the layer, droplets form, and they are drawn down by the force of gravity. *See also* Marangoni Effect.

Winsor-Type Emulsions   Several categories of microemulsions that refer to equilibrium phase behaviors and distinguish, for example, the number of phases that may be in equilibrium and the nature of the continuous phase. *See* reference 39. They are denoted as Winsor Type I (oil-in-water), Type II (water-in-oil), Type III (most of the surfactant is in a middle phase with oil and water), and Type IV (water, oil, and surfactant are all present in a single phase). The Winsor Type III system is sometimes referred to as a middle-phase microemulsion, and the Type IV system is often referred to simply as a microemulsion. An advantage of the Winsor category system is that it is independent of the density of the oil phase and may lead to less ambiguity than do the lower-phase or upper-phase microemulsion type terminology. Nelson-type emulsions are similarly identified, but with different type numbers.

W/O   Abbreviation for a water-dispersed-in-oil emulsion.

Work of Adhesion   The energy of attraction between molecules in a phase. Defined as the work per unit area done on a system when two phases meeting at an interface of unit area are separated reversibly to form unit areas of new interfaces of each with a third phase.

Work of Cohesion   The work per unit area done on a system when a body of a phase is separated reversibly to form two bodies of the phase, each forming unit areas of new interfaces with a third phase.

Work of Immersional Wetting   *See* Wetting Tension.

Work of Separation   Synonym for the work of adhesion.

Work of Spreading   Expressed per unit area, this is the same as the spreading coefficient.

W/O/A   Abbreviation for a thin fluid film of oil between water and air phases. *See also* Fluid Film.

W/O/W   In multiple emulsions: Abbreviation for a water-dispersed-in-oil-dispersed-in-water multiple emulsion. Here the oil droplets have water droplets dispersed within them, and the oil droplets themselves are dispersed in water forming the continuous phase.

In fluid films: Abbreviation for a thin fluid film of oil in a water phase. Note the possibility of confusion with the multiple emulsion convention. *See also* Fluid Film.

Xerogel   A xerogel is not a gel but rather is used with reference to a dried-out, possibly open structure that was a gel. Also spelled zerogel. Example: silica gel.

XES   *See* X-Ray Emission Spectroscopy.

XPS   X-ray photoelectron spectroscopy. *See* Photoelectron Spectroscopy.

X-Ray Diffraction (XRD)   A technique in which the scattering of X-rays by a crystal lattice is measured and used to determine the crystal's structure.

X-Ray Emission Spectroscopy (XES)   A technique used for the determination of surface composition by scanning the surface with an X-ray or electron beam. The beam ionizes surface atoms by ejecting inner-shell electrons. Electron transfer from outer electron shells will result in the emission of energy as characteristic X-rays. The spectrum of the emitted X-rays is then determined. A derived technique is soft X-ray emission spectroscopy (SXES). *See also* Auger Electron Spectroscopy, Table 9.

X-Ray Photoelectron Spectroscopy (XPS)   *See* Photoelectron Spectroscopy.

XRD   *See* X-Ray Diffraction.

Yield Stress   For some fluids, the shear rate (flow) remains at zero until a threshold shear stress is reached; this is termed the yield stress. Beyond the yield stress flow begins. Also termed the yield value.

Yield Value   *See* Yield Stress.

Young's Equation   A fundamental relationship giving the balance of forces at a point of three-phase contact. For a gas–liquid–solid system Young's equation is $\gamma_{SL} + \gamma_{LG} \cos \theta = \gamma_{SG}$, where γ_{SL}, γ_{LG}, and γ_{SG} are interfacial tensions between solid–liquid, liquid–gas, and solid–gas, respectively, and θ is the contact angle of the liquid with the solid, measured through the liquid.

Young–Laplace Equation   The fundamental relationship giving the pressure difference across a curved interface in terms of the surface or interfacial tension and the principal radii of curvature. In the special case of a spherical interface, the pressure difference is equal to twice the surface (or interfacial) tension divided by the radius of curvature. Also referred to as the equation of capillarity.

Young Modulus   *See* Hooke's Law.

Zeolites   A class of aluminosilicate minerals having large cavities in their crystal structures. These allow ion exchange of large ions and also can permit the size-selective passage of organic molecules. They are used as ion exchangers, molecular sieves, and catalysts.

Zerogel   *See* Xerogel.

Zero Point of Charge   *See* Point of Zero Charge, Electrocapillarity.

Zeta Potential   Strictly called the electrokinetic potential, the zeta potential refers to the potential drop across the mobile part of the electric double layer. Any species undergoing electrokinetic motion, such as electrophoresis, moves with a certain immobile part of the electric double layer that is assumed to be distinguished from the mobile part by a distinct plane, the shear plane. The zeta potential is the potential at that plane, and is calculated from measured electrokinetic mobilities (e.g., electrophoretic mobility) or potentials (e.g., sedimentation potential) by using one of a number of available theories.

Zimm Plot   A graph used in the determination of root-mean-square end-to-end distances of dispersed species based on light-scattering data.

Zisman Plot   *See* Critical Surface Tension of Wetting.

Zone Electrophoresis   A method for the separation of charged colloidal particles or large molecules. An electric-field gradient is imposed along a supporting medium, such as a gel, and a sample of mixture to be separated is applied to one end of the supporting medium. As electrophoretic motion occurs, regions of different components separate out along the direction of the electric-field gradient according to the different electrophoretic mobilities of the components. *See also* Isoelectric Focusing.

Zwitterionic Surfactant   A surfactant molecule that contains both negatively and positively charged groups. Example: lauramido-propylbetaine, $C_{11}H_{23}CONH(CH_2)_3N^+(CH_3)_2CH_2COO^-$, at neutral and alkaline solution pH. *See* Amphoteric Surfactant.

Appendix: Useful Tables

Table 1. Some Classifications for Atmospheric Aerosols of Liquid Droplets

Classification	Lower–Upper Size Limit (μm)
Fog	0.5–30
Cloud	2–100
Drizzle	100
Rain	100–about 5000

SOURCE: References 40 and 41.

Table 2. Types of Colloidal Dispersion

Dispersed Phase	Dispersion Medium	Name	Examples
Liquid	Gas	Aerosol of liquid droplets	Fog, mist
Solid	Gas	Aerosol of solid particles	Smoke, dust
Gas	Liquid	Foam	Soap suds
Liquid	Liquid	Emulsion	Milk, mayonnaise
Solid	Liquid	Sol, suspension	Inks, gels, bacteria in water
Gas	Solid	Solid foam	Polystyrene foam
Liquid	Solid	Solid emulsion	Opal, pearl
Solid	Solid	Solid suspension	Ruby-stained glass

Table 3.   Some Particle Size Classifications

Classification	Wentworth	Soil Sci. Soc. Amer.
Clays	0–3.9	0–2.0
Silt	3.9–62.5	2.0–50
Very fine sand	62.5–125	50–100
Fine sand	125–250	100–250
Medium sand	250–500	250–500
Coarse sand	500–1000	500–1000
Very coarse sand	1000–2000	1000–2000
Gravel/Granule	2000–4000	2000–80,000

NOTE: All values are lower and upper size limits in micrometers.

SOURCE: References 42 and 43.

Table 4.   Glossary of Viscosities

Term	Symbol	Explanation
Absolute viscosity	η	$\eta = \tau/\dot{\gamma}$ and can be traced to fundamental units independent of the type of instrument
Apparent viscosity	η_{APP}	$\eta_{APP} = \tau/\dot{\gamma}$ but as determined for a non-Newtonian fluid, usually by a method suitable only for Newtonian fluids.
Differential viscosity	η_D	$\eta_D = d\,\tau/d\,\dot{\gamma}$
Inherent viscosity	η_{Inh}	$\eta_{Inh} = C^{-1}\ln(\eta/\eta_o)$
Intrinsic viscosity	$[\eta]$	$[\eta] = \lim_{C\to 0}\lim_{\dot{\gamma}\to 0} \eta_{sp}/C$
		$[\eta] = \lim_{C\to 0}\lim_{\dot{\gamma}\to 0} (1/C)\ln \eta_{Rel}$
Kinematic viscosity	η_K	$\eta_K = \eta/\rho$
Reduced viscosity	η_{Red}	$\eta_{Red} = \eta_{sp}/C$
Relative viscosity	η_{Rel}	$\eta_{Rel} = \eta/\eta_o$
Specific increase in viscosity	η_{SP}	$\eta_{SP} = \eta_{Rel} - 1$

SYMBOLS: ρ is the density of the bulk fluid or dispersion, η_o is the viscosity of the pure solvent or dispersion medium, and C is the dispersed-phase concentration (usually volume fraction).

Table 5. Equations for Predicting Viscosities of Dispersions

Name	Equation	Explanation
Eilers equation	$\eta = \eta_0(1 + 2.5\phi + 4.94\phi^2 + 8.78\phi^3)$	Emulsions of viscous oils
Einstein equation	$\eta = \eta_0(1 + 2.5\phi)$	For $\phi < 0.02$
Guth–Gold–Simha equation	$\eta = \eta_0(1 + 2.5\phi + 14.1\phi^2)$	For $\phi < 0.06$
Hatschek equation	$\eta = \eta_0(1/\{1 - \phi^{1/3}\})$	For emulsions, $\phi > 0.5$
Mooney equation	$\eta = \eta_0 \exp(2.5\phi/\{1 - a\phi\})$	For emulsions
Oliver–Ward equation	$\eta = \eta_0(1 + a\phi + a^2\phi^2 + a^3\phi^3 + \dots)$	For spheres
Richardson equation	$\eta = \eta_0 \exp(a\phi)$	For emulsions
Sibree equation	$\eta = \eta_0(1/\{1 - (a\phi)^{1/3}\})$	For emulsions
Simha equation	$\eta = \eta_0(1 + 32\phi/\{15p\pi\})$	Anisometric particle suspensions
Staudinger–Mark–Houwink equation	$[\eta] = KM^a$	For polymer solutions
Taylor equation	$\eta = \eta_0(1 + 2.5\phi\{(\eta_D + 0.4\eta_0)/(\eta_D + 0.4\eta_0)\})$	Emulsions of viscous oils
Thomas equation	$\eta = \eta_0(1 + 2.5\phi + 10.05\phi^2 + 0.00273 \exp(16.6\phi))$	For suspensions

SYMBOLS: η_0 is the viscosity of the pure solvent or continuous phase, η_D is the viscosity of the dispersed phase, K and a are empirical constants, and p is the ratio of minor to major particle axes.

Table 6. Equations for Predicting Conductivities of Dispersions

Name	Equation	Explanation
Bruggeman equation	$(\kappa - \kappa_D)(\kappa_C/\kappa)^{1/3} = (1 - \phi)(\kappa_C - \kappa_D)$ $(\kappa/\kappa_C) = (1 - \phi)^{3/2}$	When $\kappa_C \gg \kappa_D$
Fricke equation	$\kappa = \kappa_C - [\phi(\kappa - \kappa_D)/3(1 - \phi)]\Sigma\kappa_C/$ $[\kappa_C(1 - a_i) + \kappa_D a_i]$	a_i is a series of further terms
Hanai equation	$(\kappa/\kappa_C) = 3\varepsilon(\varepsilon - \varepsilon_C)/$ $[(\varepsilon_D + 2\varepsilon)(\varepsilon_D - \varepsilon_C)]$	For high-frequency measurement
	$(\kappa/\kappa_C) = 1/(1 - \phi)^3$	For low-frequency measurement
Wagner equation	$(\kappa - \kappa_C)/(\kappa + 2\kappa_C) = \phi(\kappa_D - \kappa_C)/$ $(\kappa_D + 2\kappa_C)$	

SYMBOLS: κ is the conductivity of the dispersion, κ_D is the conductivity of the dispersed phase, κ_C is the conductivity of the continuous phase, and a is a constant.

SOURCE: Reference 13.

Table 7. Equations for Predicting Relative Permittivities of Dispersions

Name	Equation	Explanation
Böttcher equation	$(\varepsilon - \varepsilon_C)/(3\varepsilon) = \phi[(\varepsilon_D - \varepsilon_C)/(\varepsilon_D + 2\varepsilon)]$	
Bruggeman equation	$(\varepsilon - \varepsilon_D)/(\varepsilon_C - \varepsilon_D) = (1 - \phi)(\varepsilon/\varepsilon_C)^{1/3}$	For high ϕ
Fradkina equation	$\varepsilon = \varepsilon_C(1 + 3\phi)$	For W/O emulsions
Hanai equation	$(\varepsilon/\varepsilon_C) = 1/(1 - \phi)^3$	For low-frequency measurement
Wiener equation	$(\varepsilon - \varepsilon_C)/(\varepsilon + 2\varepsilon_C) =$ $\phi(\varepsilon_D - \varepsilon_C)/(\varepsilon_D + 2\varepsilon_C)$	For $\phi \ll 1$

SYMBOLS: ε is the relative permittivity of the dispersion, ε_D is the relative permittivity of the dispersed phase, ε_C is the relative permittivity of the continuous phase, and a is a constant.

SOURCE: Reference 13.

Table 8. Equations for Predicting Surface and Interfacial Tensions

Name	Equation	Explanation
Antonow's rule	$\gamma_{12} = \gamma_1 - \gamma_2$	
Eötvös equation	$\gamma^\circ = k(T_c - T)(M/\rho)^{-2/3}$	
Fowkes equation	$\gamma_{12} = \gamma_1 + \gamma_2 - 2\sqrt{(\gamma_1{}^d \gamma_2{}^d)}$	γ^d represents dispersion components of surface tensions
Girifalco–Good equation	$\gamma_{12} = \gamma_1 + \gamma_2 - 2a(\sqrt{\gamma_{12}})$	
Ideal mixing rule	$\gamma^\circ{}_s = \gamma^\circ{}_1 X_1 + \gamma^\circ{}_2 X_2$	For solutions, s, of two components, 1 and 2, having mole fractions X_1 and X_2
Sugden's parachor equation	$\gamma^\circ = (P\Delta\rho/M)^4$	P is an empirical constant, the parachor
Szyszkowski equation	$\gamma^\circ = \gamma^\circ{}_0 \{1 - b \ln((1 + C)/a)\}$	Aqueous solutions of concentration C

SYMBOLS: a and b are empirical constants, M is molecular mass, T_c is critical temperature, ρ is density, and $\gamma^\circ{}_0$ is the surface tension of pure solvent.

Table 9. Some Surface Techniques and Their Acronyms

Acronym	Technique	Information
AES	Auger electron spectroscopy	Surface composition
AEAPS	Auger electron appearance potential spectroscopy	Surface composition
AFM	Atomic force microscopy	Surface morphology
APD	Azimuthal photoelectron diffraction	Surface structure
APS	Appearance potential spectroscopy	Surface composition
CELS	Characteristic energy-loss spectroscopy	Adsorbed composition
EELS	Electron energy-loss spectroscopy	Adsorbed composition
EID	Electron-impact desorption spectroscopy	Surface, adsorbed composition
EIS	Electron-impact spectroscopy	Adsorbed composition
ELEED	Elastic low-energy electron diffraction	Surface structure
ELS	Energy-loss spectroscopy	Adsorbed composition
ESCA	Electron spectroscopy for chemical analysis	Surface composition
ESD	Electron-stimulated desorption	Surface structure, composition
EXAFS	Extended X-ray absorption fine structure spectroscopy	Surface atom packing
FDS	Flash desorption spectroscopy	Surface, adsorbed composition
FEM	Field emission microscopy	Surface structure
FIM	Field ion microscopy	Surface structure
HEED	High-energy electron diffraction	Surface structure
HEIS	High-energy ion scattering	Surface composition, structure
HREELS	High-resolution electron energy-loss spectroscopy	Surface structure, composition
ILS	Ionization-loss spectroscopy	Surface composition
INS	Ion-neutralization spectroscopy	Surface, adsorbed electron structure
IRAS	Infrared reflection–adsorption spectroscopy	Surface structure, composition
ISS	Ion-scattering spectroscopy	Surface composition
LEED	Low-energy electron diffraction	Surface structure
LEIS	Low-energy ion scattering	Surface composition, structure

Table 9.   Continued

Acronym	Technique	Information
MEIS	Medium-energy ion scattering	Surface composition, structure
MBS	Molecular beam spectroscopy	Surface reaction kinetics
NPD	Normal photoelectron diffraction	Surface structure
OSEE	Optically stimulated exoelectron emission spectroscopy	Adsorbed composition
PAS	Photoacoustic spectroscopy	Surface, adsorbed vibrational states
PES	Photoelectron spectroscopy	Surface composition
PhD	Photoelectron diffraction	Surface structure
PSD	Photon-stimulated desorption	Surface structure, composition
RHEED	Reflection high-energy electron diffraction	Surface structure, composition
SEM	Scanning electron microscopy	Surface morphology, composition
SEXAFS	Surface-extended X-ray absorption fine structure	Surface structure, composition
SHEED	Scanning high-energy electron diffraction	Surface heterogeneity
SIMS	Secondary ion mass spectroscopy	Surface composition
STM	Scanning tunneling microscopy	Surface morphology
SXAPS	Soft X-ray appearance potential spectroscopy	Surface composition
SXES	Soft X-ray emission spectroscopy	Surface composition
TDS	Thermal desorption spectroscopy	Surface, adsorbed composition
TPDS	Temperature-programmed desorption spectroscopy	Surface, adsorbed composition
TPRS	Temperature-programmed reaction spectroscopy	Surface, adsorbed composition
UPS	Ultraviolet photoelectron (photoemission) spectroscopy	Surface structure, composition
XES	X-ray emission spectroscopy	Surface composition
XPS	X-ray photoelectron spectroscopy	Surface structure, composition

SOURCE: References 10 and 44.

References

1. Kerker, M. *J. Colloid Interface Sci.* **1987,** *116(1),* 296–299.
2. Asimov, I. *Words of Science and the History Behind Them;* Riverside Press: Cambridge, MA, 1959.
3. Freundlich, H. *Colloid and Capillary Chemistry;* English translation of 3rd. ed., Methuen: London, 1926.
4. *Manual of Symbols and Terminology for Physicochemical Quantities and Units;* Appendix II; Prepared by IUPAC Commission on Colloid and Surface Chemistry; Butterworths: London, 1972. *See also* the additions in *Pure Appl. Chem.* **1983,** *55,* 931–941; *Pure Appl. Chem.* **1985,** *57,* 603–619.
5. Williams, H. R.; Meyers, C. J. *Oil and Gas Terms;* 6th ed.; Matthew Bender: New York, 1984.
6. *A Dictionary of Petroleum Terms;* 2nd ed.; Petroleum Extension Service, University of Texas at Austin: Austin, TX, 1979.
7. *The Illustrated Petroleum Reference Dictionary;* 2nd ed.; Langenkamp, R. D., Ed.; PennWell Books: Tulsa, OK, 1982.
8. *McGraw-Hill Dictionary of Scientific and Technical Terms;* 3rd. ed.; Parker, S. P., Ed.; McGraw-Hill: New York, 1984.
9. Becher, P. *Dictionary of Colloid and Surface Science;* Dekker: New York, 1990.
10. Adamson, A. W. *Physical Chemistry of Surfaces;* 5th. ed.; Wiley: New York, 1990.
11. Schramm, L. L. In *Emulsions: Fundamentals and Applications in the Petroleum Industry;* Schramm, L. L., Ed.; Advances in Chemistry Series 231; American Chemical Society: Washington, DC, 1992; pp 385–405.
12. Schramm, L. L. In *Foams: Fundamentals and Applications in the Petroleum Industry;* Schramm, L. L., Ed.; Advances in Chemistry Series 242; American Chemical Society: Washington, DC, in press.
13. Becher, P. *Emulsions: Theory and Practice;* 2nd. ed., Krieger: Malabar, FL, 1977.
14. Whorlow, R. W. *Rheological Techniques;* Ellis Horwood: Chichester, England, 1980.
15. Myers, D. *Surfactant Science and Technology;* VCH: New York, 1988.
16. Rosen, M. J. *Surfactants and Interfacial Phenomena;* Wiley: New York, 1978.
17. *Microemulsions and Emulsions in Foods;* El-Nokaly, M.; Cornell, D., Eds.; ACS Symposium Series 448, American Chemical Society: Washington, DC, 1991.
18. *Surface and Colloid Chemistry in Natural Waters and Water Treatment;* Beckett, R., Ed.; Plenum Press: New York, 1990.

19. van Olphen, H. *An Introduction to Clay Colloid Chemistry;* 2nd. ed.; Wiley-Interscience: New York, 1977.
20. Enriquez, L. G.; Flick, G. J. *Dev. Food Sci.* **1989,** *19,* 235–334.
21. Anderson, W. G. *J. Petrol. Technol.* **1986,** *38(12),* 1246–1262.
22. Isenberg, C. *The Science of Soap Films and Soap Bubbles;* Tieto Ltd.: Clevedon, United Kingdom, 1978, p 101.
23. *Glossary of Soil Science Terms;* Soil Science Society of America: Madison, WI, 1987.
24. *Glossary of Terms in Soil Science;* Research Branch, Agriculture Canada: Ottawa, Ontario; revised 1976.
25. Martinez, A. R. In *The Future of Heavy Crude and Tar Sands;* Meyer, R. F.; Wynn, J. C.; Olson, J. C., Eds.; UNITAR: New York, 1982; pp ixvii–ixviii.
26. Danyluk, M.; Galbraith, B.; Omana, R. In *The Future of Heavy Crude and Tar Sands;* Meyer, R. F.; Wynn, J. C.; Olson, J. C., Eds.; UNITAR: New York, 1982; pp 3–6.
27 Khayan, M. In *The Future of Heavy Crude and Tar Sands;* Meyer, R. F.; Wynn, J. C.; Olson, J. C., Eds.; UNITAR: New York, 1982; pp 7–11.
28. Kukla, R. J. *Chem. Eng. Progr.* **1991,** *87,* 23–35.
29. Cash, L.; Cayias, J. L.; Fournier, G.; Macallister, D.; Schares, T.; Schechter, R. S.; Wade, W. H. *J. Colloid Interface Sci.* **1977,** *59,* 39–44.
30. Cayias, J. L.; Schechter, R. S.; Wade, W. H. *Soc. Petrol. Eng. J.* **1976,** *December,* 351–357.
31. Harwell, J. H.; Hoskins, J. C.; Schechter, R. S.; Wade, W. H. *Langmuir* **1985,** *1(2),* 251–262.
32. Ross, S.; Morrison, I. D. *Colloidal Systems and Interfaces;* Wiley: New York, 1988.
33. Schramm, L. L.; Clark, B. W. *Colloids Surf.* **1983,** *7,* 135–146.
34. Nelson, R. C. *Chem. Eng. Prog.* **1989,** *March,* 50–57.
35. Becher, P. *J. Colloid Interface Sci.* **1990,** *140(1),* 300–301.
36. Kosmulski, M.; Matijevi , E. *J. Colloid Interface Sci.* **1992,** *150(1),* 291–294.
37. Capes, C. E.; McIlhinney, A. E.; Sirianni, A. F. In *Agglomeration 77;* American Institute of Mining, Metallurgical, and Petroleum Engineers: New York, NY, 1977; pp 910–930.
38. Ross, S.; Becher, P. *J. Colloid Interface Sci.* **1992,** *149,* 575–579.
39. Winsor, P. A. *Solvent Properties of Amphiphilic Compounds;* Butterworths: London, 1954.
40. Warneck, P. *Chemistry of the Natural Atmosphere;* Academic: San Diego, CA, 1988.
41. Rogers, R. R.; Yau, M. K. *A Short Course in Cloud Physics;* 3rd. ed.; Pergamon: Oxford, England, 1989.

42. Scholle, P. A. *Constituents, Textures, Cements, and Porosities of Sandstones and Associated Rocks;* Memoir 28; American Association of Petroleum Geologists: Tulsa, OK, 1979.
43. Blatt, H.; Middleton, G.; Murray, R. *Origin of Sedimentary Rocks;* 2nd. ed.; Prentice-Hall: Englewood Cliffs, NJ, 1980.
44. Woodruff, D. P.; Delchar, T. A. *Modern Techniques of Surface Science;* Cambridge University Press: New York, 1986.
45. Fuerstenau, D. W.; Williams, M. C. *Colloids Surf.* **1987,** *22,* 87–91.
46. Spelt, J. K. *Colloids Surf.* **1990,** *43,* 389–411.
47. Schramm, L. L.; Novosad, J. J. *Colloids Surf.* **1990,** *46,* 21–43.

Highlights from ACS Books

The Language of Biotechnology: A Dictionary of Terms
By John M. Walker and Michael Cox
254 pp; clothbound, ISBN 0–8412–1489–1; paperback, ISBN 0–8412–1490–5

Silent Spring Revisited
Edited by Gino J. Marco, Robert M. Hollingworth, and William Durham
214 pp; clothbound, ISBN 0–8412–0980–4; paperback, ISBN 0–8412–0981–2

Chemistry and Crime: From Sherlock Holmes to Today's Courtroom
Edited by Samuel M. Gerber
135 pp; clothbound, ISBN 0–8412–0784–4; paperback, ISBN 0–8412–0785–2

From Caveman to Chemist: Circumstances and Achievements
By Hugh W. Salzberg
300 pp; clothbound, ISBN 0–8412–1786–6; paperback, ISBN 0–8412–1787–4

The Green Flame: Surviving Government Secrecy
By Andrew Dequasie
300 pp; clothbound, ISBN 0–8412–1857–9

Trends in Chemical Consulting
Edited by Charles S. Sodano and David M. Sturmer
165 pp; paperback ISBN 0–8412–2106–5

The Basics of Technical Communicating
By B. Edward Cain
198 pp; clothbound, ISBN 0–8412–1451–4; paperback, ISBN 0–8412–1452–2

Phosphorus Chemistry in Everyday Living
By Arthur D. F. Toy and Edward N. Walsh
362 pp; clothbound, ISBN 0–8412–1002–0

Steroids Made It Possible
By Carl Djerassi
205 pp; clothbound, ISBN 0–8412–1773–4

For further information and a free catalog of ACS books, contact:
American Chemical Society
Distribution Office, Department 225
1155 16th Street, NW, Washington, DC 20036
Telephone 800–227–5558

697642